STUDENT'S GUIDE TO THE IET WIRING REGULATIONS

Published by The Institution of Engineering and Technology, London, United Kingdom

The Institution of Engineering and Technology is registered as a Charity in England & Wales (no. 211014) and Scotland (no. SC038698).

The Institution of Engineering and Technology is the institution formed by the joining together of the IEE (The Institution of Electrical Engineers) and the IIE (The Institution of Incorporated Engineers).

© 2016, 2018 The Institution of Engineering and Technology

First published 2015 (978-1-78561-078-3)
Second Edition 2018 (978-1-78561-474-3)

This publication is copyright under the Berne Convention and the Universal Copyright Convention. All rights reserved. Apart from any fair dealing for the purposes of research or private study, or criticism or review, as permitted under the Copyright, Designs and Patents Act, 1988, this publication may be reproduced, stored or transmitted, in any form or by any means, only with the prior permission in writing of the publishers, or in the case of reprographic reproduction in accordance with the terms of licences issued by the Copyright Licensing Agency. Enquiries concerning reproduction outside those terms should be sent to the publishers at The Institution of Engineering and Technology, Michael Faraday House, Six Hills Way, Stevenage, SG1 2AY, United Kingdom.

Copies of this publication may be obtained from:
The Institution of Engineering and Technology
PO Box 96, Stevenage, SG1 2SD, UK
Tel: +44 (0)1438 767328
Email: sales@theiet.org

www.electrical.theiet.org/books

While the author, publisher and contributors believe the information and guidance given in this work are correct, all parties must rely upon their own skill and judgement when making use of them. The author, publisher and contributors do not assume any liability to anyone for any loss or damage caused by any error or omission in the work, whether such an error or omission is the result of negligence or any other cause. Where reference is made to legislation it is not to be considered as legal advice. Any and all such liability is disclaimed.

ISBN 978-1-78561-474-3 (wiro bound)
ISBN 978-1-78561-475-0 (electronic)

Typeset in the UK by the Institution of Engineering and Technology
Printed in the UK by A. McLay and Company Ltd, Cardiff

Contents

List of Figures	7
List of Tables	11
About the author	12
Acknowledgements	13
Welcome	**15**
What is the right route for you?	15
What will you get from this Guide?	19
The road ahead	20

Section 1 The IET Wiring Regulations — **21**

1.1	The IET Wiring Regulations	21
1.2	Who is responsible for writing the IET Wiring Regulations?	22
1.3	IET standards and compliance staff	24
1.4	Symbols and SI units	31
1.5	Vocabulary: terms, acronyms and abbreviations	32
1.6	Policies and legal requirements	43
1.7	Guidance on pre-determined values in the IET Wiring Regulations	44

Section 2 Health and Safety — **49**

2.1	Introduction	49
2.2	Risk assessments and method statement	51
2.3	Safe isolation procedure	58
2.4	Tool safety	61

Section 3 Generation and Transmission		69
3.1	What is used to generate electricity?	69
3.2	What are the characteristics of the electricity generated in the UK?	71
3.3	How is electricity transmitted?	71
3.4	How do we measure what we use?	75

Section 4 Supply		79
4.1	Supply intake arrangements	79
4.2	Live conductors	79
4.3	Current	80
4.4	Earthing arrangement	81
4.5	Responsibility	84
4.6	Tails	85
4.7	The consumer unit	86

Section 5 Protection and Isolation		93
5.1	Protective devices	93
5.2	Selecting protective devices	107
5.3	Selectivity	109
5.4	Integration of devices and components	112
5.5	Isolation and switching (Chapter 53)	112
5.6	Isolators and switches (Chapter 46)	113
5.7	Isolation	114
5.8	Switching off for mechanical maintenance	115
5.9	Emergency switching	116
5.10	Functional switching	116
5.11	Undervoltage protection	117

Section 6 Earthing and Bonding		121
6.1	What is earthing and why do we need it?	121
6.2	Protective equipotential bonding	128
6.3	Supplementary equipotential bonding	134

Section 7 Cable Calculations — 137
7.1 Why do we need cable calculations? — 137
7.2 Current rating of equipment, protective devices and current-carrying capacity of cable — 138
7.3 How design current, protective device rating and current-carrying capacity relate to each other — 143
7.4 Voltage drop — 149

Section 8 Final circuits — 155
8.1 General installation practices — 155
8.2 Designing and installing a new circuit — 156
8.3 Luminaires and lighting installations — 163
8.4 Radial final circuits (power) — 175
8.5 Ring final circuits — 177
8.6 Spurs — 179
8.7 Shower circuit — 182
8.8 Separated Extra-Low Voltage systems (SELV) — 183
8.9 Protective Extra-Low Voltage systems (PELV) — 185
8.10 Functional Extra-Low Voltage systems (FELV) — 186

Section 9 Inspection and Testing — 189
9.1 Why test? — 189
9.2 Initial inspection — 191
9.3 The tests in detail — 193
9.4 Dead tests — 195
9.5 Live tests — 212

Section 10 Fault Finding — 219
10.1 What is a fault? — 219
10.2 How to diagnose and locate a fault — 222

Section 11 Common Calculations — 235
11.1 Simple transposition — 235
11.2 Triangle method: voltage, amps, resistance and watts — 237
11.3 Resistors — 238

Section 12 Diversity 243
12.1 What is diversity? 243
12.2 Calculating diversity 244
12.3 Diversity for ring final circuits 246

Appendix A Special Locations 251

Appendix B Table of Symbols 269

Answers 279

Section 1	The IET Wiring Regulations	279
Section 2	Health and Safety	279
Section 3	Generation and Transmission	280
Section 4	Supply	281
Section 5	Protection and Isolation	281
Section 6	Earthing and Bonding	282
Section 7	Cable calculation	283
Section 8	Final Circuits	285
Section 9	Inspection and Testing	286
Section 10	Fault Finding	286
Section 11	Common Calculations	287
Section 12	Diversity	287

Index 289

List of Figures

Figure 1.1	Extract from the *Wiring Rules and Regulations*, published 1882	22
Figure 1.2	Examples of JPEL/64 Committee membership	25
Figure 1.3	Contents page of the IET Wiring Regulations	28
Figure 1.4	Detail of a regulation number	30
Figure 1.5	Table 4A2 - IET Wiring Regulations	45
Figure 1.6	Tables 4D1A to 4E2A - IET Wiring Regulations	46
Figure 1.7	Table 4B1 - IET Wiring Regulations	47
Figure 2.1	Example of a risk assessment	53
Figure 2.2	Risk assessment flow chart	56
Figure 2.3	Test leads	59
Figure 2.4	Safe isolation procedure	60
Figure 3.1	Generation of electricity	70
Figure 3.2	Transmission from power station to National Grid	72
Figure 3.3	Examples of transmission towers	73
Figure 3.4	Examples of distribution transformers	74
Figure 3.5	Electricity meters	75
Figure 3.6	Transmission and distribution network	76
Figure 4.1a	TN-C-S arrangement	82
Figure 4.1b	TN-S arrangement	83
Figure 4.1c	TT arrangement	84
Figure 4.2	Typical intake layouts and responsibilities	85
Figure 4.3	Distribution and final circuits	88
Figure 4.4	Six-way consumer unit with eight modules	89
Figure 4.5	Split-way board with three ways protected by one RCD and three ways protected by another (12 modules in total)	90
Figure 5.1	Time/current graph as shown in Appendix 3 BS 7671	95
Figure 5.2	HRC fuse	97
Figure 5.3	Cartridge fuse	98
Figure 5.4	Plug-top fuse	98

Figure 5.5	BS 3036 Semi enclosed fuse	99
Figure 5.6	Relationship between I_{CS} and I_{CN} for circuit-breakers	101
Figure 5.7	Internal components within a circuit-breaker	102
Figure 5.8	Time current graphs for circuit-breakers in Appendix 3 of the IET Wiring Regulations	103
Figure 5.9	Typical RCD arrangement	105
Figure 5.10	Time/performance criteria for RCDs	105
Figure 5.11	Examples of discrimination	110
Figure 5.12	A typical motor circuit	118
Figure 6.1	TN-C-S arrangement	125
Figure 6.2	TN-S arrangement	126
Figure 6.3	TT arrangement	128
Figure 6.4	Installation without protective equipotential bonding	132
Figure 6.5	Installation with protective equipotential bonding	133
Figure 6.6	Unbroken bonding conductor	134
Figure 7.1	Temperature of environment	139
Figure 7.2	Cables buried in the ground	140
Figure 7.3	Depth of the cables buried in the ground	140
Figure 7.4	BS 3036 Fuses	141
Figure 7.5	Cable grouping	141
Figure 7.6	Cable run through thermal insulation	142
Figure 7.7	Cable run underground	142
Figure 7.8	Checking Table 4B1 from Appendix 4 of the IET Wiring Regulations	147
Figure 8.1	Recommended arrangement for an RCD protecting a domestic installation	159
Figure 8.2	Required heights for switches, socket-outlets, etc.	160
Figure 8.3	Permitted cable routes	162
Figure 8.4	Protective device	164
Figure 8.5	Terminals at the ceiling rose	165
Figure 8.6	Two-plate method	166
Figure 8.7	Three-plate method	167
Figure 8.8	Looping power	168
Figure 8.9	One-way light switch	169
Figure 8.10	Connection of conductors in a one-way light switch	170
Figure 8.11	Two-way light switch	171
Figure 8.12	Two-way conventional	171
Figure 8.13	Two-way conversion	172
Figure 8.14	Intermediate light switch	173
Figure 8.15	Intermediate (conventional) wiring configuration	173

Figure 8.16	Intermediate (conversion) wiring configuration	174
Figure 8.17	Even distribution of load over a ring final circuit	178
Figure 8.18	Table H2.1 from the *On-Site Guide*	179
Figure 8.19	Final ring circuit with spurs	181
Figure 8.20	Connection of cpc to a socket-outlet including earth tail	182
Figure 8.21	SELV Transformer	183
Figure 8.22	PELV Transformer	185
Figure 8.23	FELV Transformer	186
Figure 9.1	Nulling/zeroing the tester	195
Figure 9.2	Example of Method 1 on a radial final circuit	198
Figure 9.3	Example of Method 1 on a lighting circuit	199
Figure 9.4	Example of Method 2 on a radial circuit	200
Figure 9.5	Example of Method 2 on a lighting circuit	201
Figure 9.6	Testing the live conductors	203
Figure 9.7	Testing the neutral conductors	204
Figure 9.8	Testing the cpc	205
Figure 9.9	Line from leg one connected to neutral from leg two	205
Figure 9.10	Line from leg one connected to cpc from leg two	206
Figure 9.11	Testing insulation resistance between line and neutral	209
Figure 9.12	Testing insulation resistance between Neutral and Earth	210
Figure 9.13	Testing the insulation resistance between live and earth	210
Figure 9.14	Earth fault loop impedance two lead test	213
Figure 9.15	Earth fault loop impedance three lead test	214
Figure 9.16	Typical readings displayed on a phase sequence tester	216
Figure 10.1	Incorrect size terminations or connector blocks	222
Figure 10.2	Nail through cable	224
Figure 10.3	Line conductor damaged	225
Figure 10.4	Poor termination, frayed strands of conductor	226
Figure 10.5	Conductors incorrectly terminated	227
Figure 10.6	Ring final circuit, no fault	229
Figure 10.7	Ring final circuit, short circuit between line and neutral	230
Figure 10.8	Ring final circuit split into two parts	231
Figure 10.9	Ring final circuit split again	232
Figure 12.1	Extract from Table A1 *On-Site Guide*	245
Figure 12.2	Process sequence for calculating diversity for a cooker circuit	246
Figure 12.3	Ring final circuit	247
Figure 12.4	Table H2.1 from the IET's *On-Site Guide*	248
Figure 12.5	Diversity used in different circuits	249

List of Tables

Table 1.1	The Wiring Regulations standard-setters	23
Table 1.2	Types of Cable	33
Table 1.3	Institutions and organisations	38
Table 1.4	Documents published by the Health and Safety Executive	41
Table 1.5	Materials	42
Table 1.6	Expressions you might hear on site	43
Table 2.1	An everyday risk assessment: crossing the road	51
Table 2.2	Acceptable risk	54
Table 2.3	Various hand tools used by electricians	62
Table 2.4	Various power tools used by electricians	65
Table 4.1	Live conductor arrangements within the UK	80
Table 5.1	Protective devices and relevant Standards	96
Table 5.2	Protective devices	107
Table 5.3	Types of installation	114
Table 9.1	Tests for initial verification	193
Table 9.2	Minimum values of insulation resistance (Table 64 from the IET Wiring Regulations)	208
Table B.1	General symbols	269
Table B.2	International system of units: base units	271
Table B.3	Multiples and sub-multiples of quantities	271
Table B.4	SI derived units, extracted from J1.1	272
Table B.5	Symbols for use in schematic wiring diagrams	273
Table B.6	Making and breaking current	274
Table B.7	Isolating, extracted from J2.2	275
Table B.8	Making, breaking and isolating, extracted from J2.3	275
Table B.9	Meters	275
Table B.10	Location symbols for installations	276

About the author

Steven Devine,
Senior Electrical Engineer – Educational Sector

Steven Devine was recruited from the educational sector to oversee the IET's further involvement with students working with the IET Wiring Regulations. As Steven points out, when students begin studying to become an electrician, either through an apprenticeship or on a full-time course, they will generally have little or no knowledge of the IET Wiring Regulations and the related guidance, such as the *On-Site Guide* and the Guidance Note suite.

Steven's focus is to provide students with a clearer understanding of the importance of standards and regulations – who is actually responsible for them and how much voluntary time is contributed from experts from the electrotechnical industry.

Steven also serves as Secretary to Sub-Committee C Shock Protection, one of the four Sub-Committees that makes up the decision-making committee behind the IET Wiring Regulations, JPEL/64.

Acknowledgements

We would like to thank the following individuals for their significant contribution to this Guide:

Derek McGrath, Section Leader, Isle of Wight College

Peter Tanner, Lead Technical Consultant, City and Guilds

From the Institution of Engineering and Technology:

Mark Coles, Head of Technical Regulations, IET

Geoff Cronshaw, Chief Engineer, Technical Regulations, IET

Leon Markwell, Senior Engineer, Technical Regulations, IET

Amy Walker, Assistant Editor, IET

Thank you to the following organisations for their significant contribution to the images used in this Guide:

British Approvals Services for Cables (BASEC)

DeWalt

Stanley Tools

Illustrations by:

farquhar design: http://www.farquhardesign.co.uk/

Phoenix Photosetting: http://www.phoenixphotosetting.co.uk/

Welcome

If you're reading this, you've most likely decided that a career in the electrotechnical industry is for you.

Whether you're studying a full time course or you're employed as an apprentice, you've just taken your first step towards a wonderful career in the electrotechnical industry. Electricians are always in high demand, for everything from small household jobs, such as installing light fittings and changing of socket-outlets, through to commercial and industrial electrical installations. It all starts here and now.

Once you've completed an apprenticeship, or achieved the equivalent qualifications and experience, you will be regarded as an electrician. There are various organisations that provide guidance on what category of electrician you will be. This will depend primarily on your experience and qualifications.

What is the right route for you?

Below we have listed some of the various steps that can be taken to help you get into the electrotechnical industry and the possibilities that lie ahead.

(a) Full-time courses

Since the introduction of government funding for full-time electrical courses up to Level 3, there has been an increased number of courses available to school leavers. There are also basic courses available for those students who need an introduction into the construction industry. Those who have passed the required Maths and English assessments can enrol immediately on to a Level 2 course. On successful completion of a Level 2 course, students will have the opportunity to progress on to Level 3. At this stage work experience with a practising electrician is essential in order to obtain the necessary practical experience.

(b) Apprenticeships

Note: some employers require applicants to, successfully, complete a colour deficiency test as well as Maths and English assessments prior to commencement of employment.

The industry has its ups and its downs and there are times when apprenticeships are hard to come by. Luckily, now and again, the government offers financial incentives to employers when they employ an apprentice. Learning providers and local colleges are some of the places you can find out more information on funding and government incentives. It is important to be aware of these incentives when applying to companies for a job, because they do change and not all employers are aware of them.

Prior to enrolling on an apprenticeship you will need to be aware of the entry requirements. At the time this guide was published, it is expected that applicants will have at least a Grade 4 (C) GCSE or equivalent in Maths and English. Some colleges will allow their learners to obtain these qualifications during their apprenticeship. This may be quite challenging to some learners as the Maths and English qualifications will be in addition to the electrical apprenticeship. It is common for colleges and training providers to assess applicants so that they can advise whether or not they are ready to join the electrical course.

Apprenticeships are slightly different depending on the type of work you do, your employer and most importantly where you are. In England, Wales and Northern Ireland an apprentice will be working towards a Level 3 Diploma. This will usually involve attending college one day a week during term time and working with an electrician for the rest of week. In Scotland you will be attending college on block release, which means that you will be at college for a week or more at a time to achieve an SCQA Level 7 qualification. Regardless of where you are studying you will be expected to successfully complete an assessment of competence on completion of your apprenticeship. In England, Wales and Northern Ireland the assessment of competence is called the Achievement Measurement 2 (AM2-S) and in Scotland it is called the Final Integrated Competence Assessment FICA.

The apprenticeship route is ideal if you are a school leaver and have some financial support. But what if you are an adult and looking for a career change. In England and Wales there are a number of options available, Current information on the options available can be obtained from training centres, colleges and online.

In Scotland, as an adult who is new to the electrical industry there is only one route to pursue and that is the adult apprenticeship. Applicants will only be considered if they have had at least one year experience working in the electrical industry, and are employed for the duration of the apprenticeship. Adult apprentices will be expected to attend college on day release (one day every two weeks) and it is expected that they will complete their qualifications within four years of enrolment.

Once you have been accepted onto an apprenticeship you are well on your way to becoming an electrician. The qualifications you will gain during your apprenticeship will provide you with enough credits to claim what is known as a qualification framework. Once you have this, it can be submitted to various organisations, such as the Joint Industry Board (JIB) or in Scotland the Scottish Joint Industry Board (SJIB) and you can apply for your JIB/SJIB grading card, which is recognised throughout Great Britain. see the links below for more information on how to become registered:
www.jib.org.uk
www.sjib.org.uk

(c) Inspection and testing

So you have now completed either an apprenticeship or gained the necessary qualifications to be regarded as an electrician. What now?

It doesn't stop there; at this stage you have just passed your first milestone in your career progression and should now be aiming to achieve a qualification in inspection and testing. Most modern apprenticeships now incorporate a mechanism so that learners will automatically gain a qualification for inspection and testing. However, if you are currently enrolled on an apprenticeship it may be worthwhile checking that this applies to your course. If you have taken an alternative route such as, what is sometimes referred to in the industry as the improver route, you may find that you will need to complete this qualification separately.

By now, your understanding of installing wiring systems should be sufficient enough that you can work unsupervised and carry out most types of electrical installation work with confidence.

You can then move on to improve your ability to inspect work carried out by other electricians as well as to assess existing installations that may be in need of maintenance or upgrading. There are inspection and testing courses available from various awarding organisations. Most people stick with one particular awarding organisation, although this is not required. Every qualification must meet a required national standard so that they are of the same value, regardless of the awarding organisation. This essentially means that a Level 3 electrical qualification from one awarding organisation is of the same value as a Level 3 electrical qualification from any other awarding organisation.

Qualifications for inspection and testing will provide more career opportunities and introduce electricians to a supervisory/assessment role.

(d) Design

Electricians will, in many cases, be working to, and installing, someone else's design. Designers are more commonly electrical engineers who have specifically studied to become an electrical installation designer. However, electricians who have experience as installers are sought after in the design sector due to their practical understanding of electrical systems. This may involve more studying and obtaining qualifications in electrical design.

(e) Higher education

It is not uncommon for electricians to continue their education, gain qualifications such as Higher National Certificates and Diplomas (HNCs and HNDs) and continue to achieve a degree in electrical engineering or similar fields.

(f) What next? The world is your oyster...

By the time you reach this stage of your career in the electrotechnical industry you will most likely be aware of the opportunities that are available to you.

To put it simply ... there are lots!

The electrotechnical industry is vast – and growing faster than ever. Technology is fascinating, with the introduction of electric vehicles, wind generation and many other ways of harvesting our renewable energy sources, as well as smart home installations, smart cities etc. It's a fantastic time for the electrotechnical industry at the height of technological advancement.

(g) Around the world

It is important to remember that a British qualification is recognised in many countries around the world. Popular destinations for qualified electricians are places such as Australia, New Zealand and Canada. Although they have slightly different standards, the principles remain the same.

What will you get from this Guide?

This Guide will provide you with information and references for various aspects of your qualification, as well as techniques that you will require for a successful career in the electrotechnical industry. Having this Guide by your side will enable you to use it to identify terminologies quickly and to see what sections of the IET Wiring Regulations (BS 7671) you need to look through in order to extract the information you need.

We've written this Guide with the aim of helping you understand the IET Wiring Regulations from a student's perspective. The Guide offers clear explanations and examples of how the IET Wiring

Regulations form an integral part of the electrotechnical industry and why they should be complied with.

Over and above this, it provides guidance on some situations electricians may come across in their work activities, such as fault finding, risk assessments and legal requirements.

Using this Guide alongside the IET Wiring Regulations and the IET's *On-Site Guide* (a book that provides installer's guidance to the IET Wiring Regulations) will provide the guidance needed while studying for a career in the electrotechnical industry.

The road ahead

Now that you know what possible career opportunities lay ahead, why not make a plan. It is never too early to start.

Career Plan	
Stage 1	For example: student/apprentice ... gain experience and knowledge in the electrotechnical industry. Work with electricians to develop skills and understanding.
Stage 2	Achieve qualifications and relevant vocational/practical skills to become an electrician.
Stage 3	Gain additional experience in the industry, offer guidance to apprentices and develop supervisory skills. Gain additional qualifications for inspection and testing and initial verification.
Stage 4	Take on supervisory roles and begin to develop management and design skills, gain qualifications for electrical design, possibly HNC.
Stage 5	Look at opportunities elsewhere, for example, other companies, other parts of the country, abroad.
Stage 6	Take on managerial roles, design installations, co-ordinate projects.
Stage 7	Continue to develop qualifications through CPD, investigate new and emerging areas and technologies.

The IET Wiring Regulations 1

After this section, you will have awareness of the following topics:

- ▶ Who is responsible for the IET Wiring Regulations?
- ▶ Information and guidance relating to the IET Wiring Regulations
- ▶ Finding your way around the IET Wiring Regulations
- ▶ Terminologies, abbreviations and acronyms
- ▶ Guidance on legal requirements
- ▶ Symbols and SI units
- ▶ Guidance on pre-determined values in the IET Wiring Regulations

This section provides information to students to enhance their understanding of the IET Wiring Regulations as well as to identify the relevance of various qualifications in the electrotechnical industry.

1.1 The IET Wiring Regulations

1.1.1 What are the IET Wiring Regulations and where do they come from?

In 1882 *The Wiring Rules and Regulations* were first issued by the Society of Telegraph Engineers and of Electricians, and consisted of four pages and 21 regulations.

▼ **Figure 1.1** Extract from the *Wiring Rules and Regulations*, published 1882

1.2 Who is responsible for writing the IET Wiring Regulations?

BS 7671 *Requirements for Electrical Installations* or, as most people call it, the IET Wiring Regulations, is a collection of regulations that are mostly made up of requirements from European standards, themselves based on international or world standards. There are slight differences in each of these standards due to varying conditions of electrical supply and installation characteristics in different countries. Table 1.1 shows which organisation is responsible for each set of standards.

▼ **Table 1.1** The Wiring Regulations standard-setters

	The world electrotechnical standard-setting body is the International Electrotechnical Commission (the IEC). The Standard is IEC 60364 Low voltage electrical installations. Any Standards introduced by IEC must usually be adopted by CENELEC and subsequently JPEL/64 within 3 years of being published.
	The European Standards body is the European Committee for Electrotechnical Standardization (CENELEC). The Standard is HD 60364 *Low voltage electrical installations*.
	The British standard-setting body is the British Standards Institution (BSI). The standard is BS 7671 and its title is *Requirements for Electrical Installations* or, more commonly, the IET Wiring Regulations. You will see on your cover of the IET Wiring Regulations that the BS 7671 is followed by a number (for example, ':2018'). This refers to the edition of the Wiring Regulations (so, for example, the 18th edition that was published in 2018).

1.3 IET standards and compliance staff

In the IET Wiring Regulations, on the page titled 'Publication Information', you will see a list of names under the heading **IET Standards and Compliance Staff**. These are the people who work directly for the IET and who manage the JPEL/64 Committee and subsidiary panels.

The JPEL/64 Committee

J stands for Joint (as in, the Committee is jointly run by the British Standards Institution and the Institution of Engineering and Technology).

P stands for Power (as in 'power generation').

EL stands for Electrical (as in 'the electrical industry').

64 is the number allocated by IEC to electrotechnical committees.

The responsibility of the Secretary

The Secretary has a key role when it comes to the work of the Committee. Before the meeting, the Secretary must:

(a) arrange a venue in which Committee members and electrical industry experts can meet.
(b) communicate with industry experts and invite new members on to the Committee.
(c) circulate agendas and the relevant documents to the Committee members – if, for example, there are new electrical products available on the market that may have an impact on the electrical industry, these should be considered by the Committee experts; and
(d) coordinate the documents to the relevant Committee members.

During the meeting, the Secretary will make a record of everything that has been discussed during the meeting, ensuring that the discussions are recorded accurately and any comments made by Committee members are logged. It is important that the Secretary can make available upon request any documents that are pertinent to the discussion.

After the meeting, the Secretary is required to prepare and circulate details of the discussions in the form of minutes to all Committee members. If there are any decisions that need to be communicated with other institutions or organisations the Secretary has the responsibility to make sure that the information is passed on. In addition, the Secretary will have a technical understanding of the new topics and knowledge of technical discussions and agreements made at previous meetings.

Key organisations on the JPEL/64 Committee

A few pages further and you will see the heading **Joint IET/BSI Technical Committees JPEL/64 Constitution**. Beneath this is a list of all the JPEL/64 Committee members and the organisation(s) they represent.

Figure 1.2 shows some of the key organisations represented at the JPEL/64 Committee.

▼ **Figure 1.2** Examples of JPEL/64 Committee membership

1.3.1 Who are the Committee members and what do they do?

Committee members are carefully selected based on their area of expertise and the organisation that they represent. It is important that every point of view is considered before introducing a new regulation or modifying an existing one.

Many of the people responsible for producing the IET Wiring Regulations have worked in the electrotechnical industry and have exceptional knowledge of electrical principles and wiring systems. These include: representatives from the Health and Safety Executive (HSE), manufacturers of electrical equipment, experts from the competent persons scheme providers and many more.

Most JPEL/64 members are sponsored and supported by their employer or trade organisation and have a passion not only to ensure that the Wiring Regulations for electrical installations meet the needs for users of electrical installations, but also to support electrical contractors to maintain practical applications of wiring systems.

1.3.2 Information and guidance for qualifications

The IET Wiring Regulations are regularly updated to meet the demands of ever improving technology, the need to use our energy as efficiently as possible and, most importantly, to protect persons and livestock from the effects of electric shock.

There are new electrical products entering the market all the time. If there are no standards set out for those products, they may perform poorly and, potentially, present a risk to the safety of the user. This also applies to the wiring systems that are designed to supply these products.

While studying for any qualification in the electrotechnical industry, part of the curriculum will involve the use of the IET Wiring Regulations. Studies that will reference the IET Wiring Regulations might stretch from basic domestic installer courses through to an Honours degree in electrical engineering. More recently, most construction qualifications, such as those on plumbing and heating and ventilation, also reference the IET Wiring Regulations.

While studying to become an electrician, students are required to have a good understanding of the IET Wiring Regulations and how to put them into practice. Regular assessments are carried out throughout courses to confirm that learners are able to understand and interpret the IET Wiring Regulations. And, when it comes to exam time, it is essential that learners have a copy of the IET Wiring Regulations and the *On-Site Guide*, as these are usually permitted for use during examination by the awarding bodies.

1.3.3 Finding your way around the IET Wiring Regulations

When you first open the IET Wiring Regulations you're bombarded with numbers and text. It can be difficult to make any sense of these and even more difficult to source the information that is needed. In this section, the basic structure of the content is explained.

Parts and appendices

The Regulations are divided into 7 Parts and 17 Appendices. Each Part is made up of Chapters and Sections, which are made up of the Regulations. The Appendices, which are included at the back of the book, are used to provide additional detail relating to the Regulations. Details of these Parts and their Chapters can be found in the contents page at the front of the IET Wiring Regulations (see Figure 1.3).

▼ **Figure 1.3** Contents page of the IET Wiring Regulations

You can familiarise yourself with the Parts to begin with:

Parts

1. Scope
2. Definitions
3. Assessment of general characteristics
4. Protection for safety
5. Selection and erection of equipment
6. Inspection and testing
7. Special installations or locations (this part of the IET Wiring Regulations has been divided into Sections not Chapters)

Next you can look at the Appendices:

Appendices

1 British Standards to which reference is made in the Regulations includes minor changes
2 Statutory regulations and associated memoranda
3 Time/current characteristics of overcurrent protective devices and RCDs
4 Current-carrying capacity and voltage drop for cables
5 Classification of external influences
6 Model forms for certification and reporting
7 Harmonized cable core colours
8 Current-carrying capacity and voltage drop for busbar trunking and powertrack systems
9 Definitions – multiple source, DC and other systems
10 Protection of conductors in parallel against overcurrent
11 Not used
12 Not used
13 Methods for measuring the insulation resistance/impedance of floors and walls
14 Determination of prospective fault current
15 Ring final circuit arrangements, Regulation 433.1
16 Devices for protection against overcurrent
17 Energy efficiency

Regulations

Now we can look at an individual regulation. This regulation number has been broken down into its relevant categories (Figure 1.4).

▼ **Figure 1.4** Detail of a regulation number

514.5.2 Protective condutor

514.5.2

5 indicates the part number. Selection and Erection.

5**1**4.5.2

5**1** indicates the chapter number. Common rules.

51**4**.5.2

51**4** indicates the section. Identification and notices.

514.**5**.2

5 indicates the group. Indentification of conductors by letters and/or numbers.

514.5.**2**

2 indicates the regulation number. <u>Protective conductor</u> Conductors with green-and-yellow colour identification shall not be numbered other than for the purpose of circuit identification.

You will notice a many regulations that end with a 200 number. For example 201, 202, 203 and so on. When a regulations ends with a 200 number it means that it is a UK specific regulation and has not been adopted from a European standard.

Test your knowledge
Have a look at the following regulations and list the list the part, chapter, section and regulation:
421.1.201
Part:
Chapter:
Section:
Regulation:

1.4 Symbols and SI units

Annex A sets out symbols and international system of units (SI units). It is important to be familiar with these symbols and units when preparing to work in the electrotechnical industry. It is also important that drawings are easy to understand and that includes the symbols that are used. Where no standard symbol is available, a specialist symbol can be used.

When symbols are used in technical drawings, they should be explained in a key or legend. In particular, a more detailed legend specifying areas such as lighting design, the type of fitting, manufacturer source output and catalogue reference number of products might be required to assist the reader to understand.

The symbols and SI units included in Annex A can also be found in the *On-Site Guide*. They are extracted mainly from IEC 60617, *Graphical symbols for diagrams*, supplemented by references from other Standards.

1.5 Vocabulary: terms, acronyms and abbreviations

Various descriptions and names are shortened, quite simply, to speed up the process of communication. If everything was spoken or written in full it would take considerably longer to get the information across. Remember:

(a) terminology can vary internationally and even from town to town. However, when it comes to engineering terminology, it is, in the most part, standardised across the UK.
(b) abbreviations are generally shortened words and, in some cases, the combination and shortening of two or more words.
(c) acronyms are the abbreviation formed from the initial letters of other words and pronounced as a word.

Whether studying electrical qualifications or working in the electrotechnical industry, you will come across a great number of phrases, terminologies, acronyms and abbreviations.

It's important to familiarise yourself with as many of these as possible. Doing so will provide you with the ability to interpret the IET Wiring Regulations and the various related publications quickly and easily.

1.5.1 Common terms

Whilst studying an electrotechnical qualification, Table 1.2 will be useful for reference.

▼ **Table 1.2** Types of cable (Images courtesy of BASEC)

ADSL	ADSL, or asymmetric digital subscriber line, is generally used for transmitting data.
CAT 5,6,7...	Category 5,6,7... generally used for computer networks such as Ethernet and data transfer.
Conductor	The essential part of the cable that carries electrical current. It is generally made of copper but can also be made of aluminium, gold and various other metals including alloys.
cpc	cpc, or circuit protective conductor, can be a conductor or cable and is designed to carry current in the event of a fault, i.e. line to exposed- or extraneous-conductive-part.
Flex (flexible)	Flexible cable, often used for portable appliances.

FP200		FP200 is sometimes considered as an alternative to MIMS. Commonly used for the installation of fire and smoke detection systems due to its high resistance to heat.
Insulation/ insulator		A layer of non-conducting material that surrounds the conductor to prevent electrical current leaking.
MICC MIMS PYRO		MICC is a mineral insulated copper clad cable and MIMS is a mineral insulated metal sheathed cable. PYRO is a brand name that is commonly used to refer to MICC and MIMS and is essentially the same, only differently named.
Multi-core		This is a cable that consists of (generally) more than two conductive cores excluding the cpc.
Sheathing inner/outer	Outer sheath, armour, Inner sheath, insulation, conductor	A protective layer that surrounds the insulation. It is designed to offer protection from mechanical damage (knocks and bumps), and is commonly made from PVC.
Single(s)		Cable(s) that consists of one conductor. It is ordinarily made up of twisted strands protected by an insulator.

SWA	SWA, or steel wire armour, is a cable that has additional steel wire strands twisted around the sheathing to provide protection from tougher impacts. The cable is commonly used outside or buried in the ground.
SY **YY** **CY**	This type of cable has a light braided steel armour that provides protection from light mechanical impact. The 'S' refers to steel braid and the 'Y' refers to PVC. YY is usually grey PVC sheath with PVC insulated flexible conductors. CY is tinned copper wire braid, usually grey PVC sheath with PVC insulated flexible conductors. For the purpose of this guidance and compliance with BS 7671, the use of non-standard cable such as SY, CY and YY is discouraged.
T&E	T&E, or flat twin & earth, PVC flat profile, two-core & earth or sometimes referred to as just twin, is a flat PVC cable that consists of three conductors. Two of the conductors have insulation (brown and blue) and one conductor has no insulation, and is designed to be used as a cpc.
Thermoplastic (PVC)	PVC, or poly-vinyl chloride, is a material that is commonly used to insulate and sheath electrical cable. This cable comes in many different permutations and conductor sizes.

Thermosetting XLPE		XLPE, or cross-linked polyethylene, is a material used to insulate and sheath cable that has a high operating temperature, e.g. 90 °C. This cable comes in many different permutations and conductor sizes.
Conductive parts		
Exposed-conductive-part		A conductive material that is part of an electrical installation that is not intended to be live, but which may become live in the event of a fault, for example, a metal case of a light fitting or the metal casing of a kettle.
Extraneous-conductive-part		A conductive material that does not form part of the electrical installation but that may become live in the event of a fault, for example, structural steel, which is in contact with the general mass of the Earth.

	Extra low voltage
ELV	ELV, or extra low voltage, is defined in the IET Wiring Regulations as a nominal voltage not exceeding 50 V AC or 120 V ripple-free DC, whether between conductors or to Earth.
FELV	FELV, or functional extra low voltage, is any other extra-low-voltage circuit that does not meet the requirements for a SELV or PELV circuit. Although the FELV part of a circuit uses an extra-low voltage, it is not protected from accidental contact with higher voltages in other parts of the circuit.
PELV	Unlike SELV, a PELV (protective extra low voltage) circuit can have a protective earth. A PELV circuit, as with SELV, should be designed to guarantee a low risk of accidental contact with a higher voltage. For a transformer, this means that the primary and secondary windings must be separated by an extra insulation barrier.
SELV	SELV, or separated extra low voltage, is an extra low voltage supply (less than 50 V AC) taken from a safe source, for example, a class II isolating transformer that complies with BS EN 61558-2-6.

1.5.2 Institutions and organisations

A lot of work goes on behind the scenes in the electrotechnical industry, much of which is achieved by the following institutions and organisations. It's useful to know how these institutions and organisations are referred to.

▼ **Table 1.3** Institutions and organisations

BS 7671	British Standard (BS) 7671 *Requirements for Electrical Installations.* Published jointly by the IET and BSI and more commonly referred to as the IET Wiring Regulations.
BS EN	British Standard & European Normalisation (Standardisation): an EN that has been implemented in the UK.
BSI	British Standards Institution. BSI is an independent national body responsible for preparing British standards.
CENELEC	French: Comité Européen de Normalisation Électrotechnique. In English this is the European Committee for Electrotechnical Standardisation.
HSE	Health and Safety Executive. The national independent watchdog for work-related health. The HSE acts in the interest of the general public to reduce work-related death and serious injury in Great Britain.

IEC		International Electrotechnical Commission. The IEC is an international standards and conformity assessment body for all fields of electrotechnology. IEC standards range from power generation, transmission and distribution to home appliances and office equipment and almost anything that forms part of an electrical system.
IET		Institution of Engineering and Technology. The IET was formed by the joining of the IEE (Institution of Electrical Engineers) and the IIE (Institution of Incorporated Engineers).
ISO		International Organization for Standardization. ISO is a voluntary organisation whose members are recognized authorities on standards, each one representing one country. ISO develop standards ranging from manufactured products and technology to food safety, agriculture and healthcare.
JPEL/64		The decision-making body for BS 7671. See Section 1.3 for more details.
ECA		Electrical Contractors Association (ECA) is the UK's largest trade association representing electrotechnical and engineering services organisations, at regional, national and European level.
SELECT		SELECT is the trade association for the electrical contracting industry in Scotland. Founded in 1900 as The Electrical Contractors' Association of Scotland.
Awarding organisations		
C&G		City and Guilds

EAL	Excellence Achievement and Learning
Competent Person Scheme providers	
ELECSA	ELECSA operates competent person schemes on behalf of the government to ensure that contractors are technically competent to carry out the work in accordance with the required regulations.
NAPIT	National Association of Professional Inspectors and Testers. NAPIT is one of the leading government-approved and United Kingdom Accreditation Service (UKAS) accredited membership scheme operators in the building services and fabric sector.
NICEIC	NICEIC is a voluntary regulatory body for the electrical contracting industry. It has been assessing the electrical competence of electricians for nearly 60 years. It currently maintains a roll of over 26,000 registered contractors and provides certification to firms across the heating, plumbing, renewable and insulation sectors.
STROMA	Stroma Certification is a provider of approved certification and training to the construction & energy industries that certify assessors, installers and organisations.

1.5.3 Acts and regulations

During your studies or when you're working as an electrician, you might come across these acts and regulations.

▼ Table 1.4 Documents published by the Health and Safety Executive

COSHH 2002	Control of Substances Hazardous to Health. COSHH ensures that adequate training is provided to staff about the safe storage, handling and use of potentially harmful chemicals.
EAWR 1989	Electricity at Work Regulations (EAWR). The EAWR ensures that working conditions in buildings do not put employees at risk of electric shock.
ESQCR 2002	Electricity Safety, Quality and Continuity Regulations 2002, as amended (ESQCR). The ESQCR is a legal document known as a Statutory Regulation (but more formally called a 'Statutory Instrument' – its number is S.I. 2002/2665). It sets out the requirements that suppliers of electricity must follow, ensuring that the electrical supply is safe and adequate for the supply requirements.
HASAWA 1974	Health & Safety at Work Act 1974. The HASAWA makes further provision for securing the health, safety and welfare of persons at work and for protecting others against risks to health or safety in connection with the activities of persons at work.
MHOR 1992	Manual Handling Operations Regulations. The employer's duty is to avoid manual handling as far as reasonably practicable if there is a possibility of injury.
PPEWR 1992	Personal Protective Equipment at Work Regulations. The responsibility lies with the employer to assess what personal protective equipment (PPE) is required for the job. The responsibility to wear such PPE lies with the employee.
PUWER 1998	Provision and Use of Work Equipment Regulations. PUWER covers the use of tools and equipment used for work activities.

RIDDOR 2013	Reporting of Injuries, Diseases and Dangerous Occurrences Regulations 2013.
	RIDDOR puts duties on employers, the self-employed and people in control of work premises (the 'responsible person') to report certain serious workplace accidents, occupational diseases and specified dangerous occurrences (near misses).

▼ **Table 1.5** Materials

Cable ladder (sometimes referred to as just a 'ladder')	As it sounds, a length of plastic or metal material that looks like a ladder, which is normally used to provide support for heavy cables **Note:** Cable tray and ladder systems are not considered to be containment as they don't completely cover or contain the cables.
Cable tray (sometimes referred to as just a 'tray')	A tray is made up of perforated sheets of metal or plastic with turned-up sides and is used more commonly in commercial and industrial environments.
Conduit	A metal or plastic pipe that commonly ranges from 16 mm to 32 mm in diameter and usually comes in 3-metre lengths. A diameter of up to 63 mm is available for larger installations. Often generally referred to as 'containment'.
Trunking	Trunking is a metal or plastic U-shaped compartment with a removable lid that is used to contain cables. Trunking can come in a wide variety of sizes and with multicompartments. Often generally referred to as 'containment'.

▼ **Table 1.6** Expressions you might hear on site

Cable pulling	When installing a cable, installers may work together to pull a very long length of cable into a building to ensure that it is pulled through the building safely without being damaged.
Chasing	When making a channel in a surface, normally brick or plaster, so that a cable can be installed and plastered over.
De-energised	An item of equipment or part of an electrical installation that has been switched off or isolated.
First fix	The first of two stages of installing a wiring system. This stage will involve preparing and installing containment for cable, pulling cables into the containment, fixing or screwing boxes for light switches, socket-outlets etc. to walls.
Fixing	The action of securing a piece of equipment to a wall or ceiling, for example, securing a conduit saddle to a wall using a screw and wall plug.
Glanding	The action of connecting a cable, such as an SWA, to an accessory. Glands are purpose made for a particular type, size and/or shape of cable and are not usually interchangeable.
Reaming	When a hole or length of conduit has been cut, the excess sharp burrs or pieces of metal or any other remnant material must be removed. This process is called reaming and the tool is called a reamer.
Second fix	The second of two stages of installing a wiring system. This stage will involve terminating electrical equipment and accessories and fitting them into their finished location. The second fix is usually carried out in the final stages of construction.
Stripping out	The removal of redundant electrical installations.

1.6 Policies and legal requirements

The IET Wiring Regulations are non-statutory, which means that they are not law. They may, however, be used in a court of law in evidence to claim compliance with statutory requirements. The relevant statutory provisions are listed in Appendix 2 of the IET Wiring Regulations and include Acts of Parliament and the regulations that those Acts have introduced. In some cases, statutory regulations may be accompanied by Codes of Practice approved under Section 16 of the Health and Safety at Work etc. Act 1974. In the

Memorandum of guidance on the Electricity at Work Regulations 1989 (HSR25), it states that "BS 7671 is a code of practice which is widely recognised and accepted in the UK and compliance with it is likely to achieve compliance with relevant aspects of the Electricity at Work Regulations 1989".

ESQCR sets out the parameters within which suppliers of electricity have to operate. These include frequency, voltage and provision of an earth.

1.7 Guidance on pre-determined values in the IET Wiring Regulations

As you begin your career in the electrical industry, you'll realise that having a good understanding of maths is essential. During your time at college you will be required to perform calculation after calculation – long equations that don't seem to have any relevance to the work that you're doing, however, it is important to understand these calculations. Fortunately, the IET and the electrotechnical committees have done a lot of the hard work for you.

When looking through the IET Wiring Regulations you will encounter lots of pre-determined values that may not make a lot of sense: tables full of correction factors, maximum earth fault loop impedance values, volt drop, grouping factors and many more.

These values have been accurately calculated and made available in the IET Wiring Regulations and related guidance to help electricians and designers in the electrotechnical industry to standardise their work and maintain safe wiring systems and working practices.

In this section there are some examples to help you understand the importance of accurate values when designing an electrical system.

1.7.1 Installation methods

Table 4A2 from the IET Wiring Regulations shows, with the aid of diagrams, the various installation methods that can be used and the correct values to use when working out current-carrying capacity of cables that are installed in a certain way.

There are over one hundred examples like the one below. Notice the reference method for this type of installation is 'C'. You'll need to remember this later when doing cable calculations.

▼ **Figure 1.5** Table 4A2 – IET Wiring Regulations

Installation Method			Reference Method to be used to determine current-carrying capacity
Number	Examples	Description	
21		Single-core or multicore cables: - fixed directly under a wooden or masonry ceiling	C (Higher than standard ambient temperatures may occur with this installation method)

1.7.2 Current-carrying capacity of conductors

The following example shows the predetermined current-carrying capacity values for cables installed using various methods. To determine these values takes a considerable amount of time and involves a lot of calculation, so would therefore place additional cost on the customer. You would have to spend a long time looking at the various factors, such as how the resistance of a conductor is affected by temperature; the trunking; whether it is installed through or alongside thermal insulation; whether or not it is a single-phase or three-phase circuit; whether it is AC or DC; etc.

This is very difficult to do and time consuming so the IET provides tables that already have the information required to design an installation.

Tables 4D1A to 4E2A of the IET Wiring Regulations show the current-carrying capacity of various types and sizes of cable using different installation methods.

Figure 1.6 Tables 4D1A to 4E2A – IET Wiring Regulations

TABLE 4D1A – Single-core 70 °C thermoplastic insulated cables, non-armoured, with or without sheath
(COPPER CONDUCTORS)

Ambient temperature: 30 °C
Conductor operating temperature: 70 °C
CURRENT-CARRYING CAPACITY (amperes):

Conductor cross-sectional area	Reference Method A (enclosed in conduit in thermally insulating wall etc.)		Reference Method B (enclosed in conduit on a wall or in trunking etc.)		Reference Method C (clipped direct)		Reference Method F (in free air or on a perforated cable tray horizontal or vertical)				
							Touching			Spaced by one diameter	
	2 cables, single-phase a.c. or d.c.	3 or 4 cables, three-phase a.c.	2 cables, single-phase a.c. or d.c.	3 or 4 cables, three-phase a.c.	2 cables, single-phase a.c. or d.c. flat and touching	3 or 4 cables, three-phase a.c. flat and touching or trefoil	2 cables, single-phase a.c. or d.c. flat	3 cables, three-phase a.c. flat	3 cables, three-phase a.c. trefoil	2 cables, single-phase a.c. or 3 cables three-phase a.c. flat	
										Horizontal	Vertical
1	2	3	4	5	6	7	8	9	10	11	12
(mm²)	(A)	(A)	(A)	(A)	(A)	(A)	(A)	(A)	(A)	(A)	(A)
1	11	10.5	13.5	12	15.5	14	-	-	-	-	-
1.5	14.5	13.5	17.5	15.5	20	18	-	-	-	-	-

The reason these values change depending on the method of installation is down to the basic principles of electricity. When current passes through a conductor it generates heat. When conductors get hotter, the resistance of the conducting material, i.e. copper, increases. If that conductor is contained in thermal insulation it will retain the heat and gradually get hotter and will therefore have less current-carrying capacity than it would if it were hanging free in the air. When looking at Tables 4D1A to 4E2A you can see that the least enclosed cables have a higher current-carrying capacity than the same cables enclosed in thermal insulation.

1.7.3 Factors

When calculating the correct size of cable to use for a particular installation, it is essential that the tables provided in the IET Wiring Regulations and other relevant guidance are used. These values are factors, most of which are based on the thermal effects of certain types of installation methods.

Some examples of tables within the IET Wiring Regulations that list factors are shown below.

▼ **Figure 1.7** Table 4B1 – IET Wiring Regulations

TABLE 4B1 – Rating factors (C_a) for ambient air temperatures other than 30 °C

Ambient temperature[a] °C	Insulation				
	60 °C thermosetting	70 °C thermoplastic	90 °C thermosetting	Mineral[a]	
				Thermoplastic covered or bare and exposed to touch 70 °C	Bare and not exposed to touch 105 °C
25	1.04	1.03	1.02	1.07	1.04
30	1.00	1.00	1.00	1.00	1.00
35	0.91	0.94	0.96	0.93	0.96

There are many more tables in the IET Wiring Regulations that provide factors for various different types of equipment and installation methods.

It is important to remember that there are a number of factors that need to be considered for any electrical installation. The IET Wiring Regulations and associated guidance provide electricians with the information they need to design and install safe and efficient wiring systems.

1.7.4 Earth fault loop impedance (EFLI)

Once an installation has been completed there are various tests that must be carried out before the circuit is energised to ensure that the installation meets the requirements of the IET Wiring Regulations. The committee of experts (JPEL/64) have calculated the impedance values that need to be achieved to ensure that protective devices disconnect within the required times given in Table 41.1 of The Wiring Regulations (more information on this can be found in testing and inspection and earthing and bonding sections).

	Test your knowledge
1.	When were wiring regulations first issued?
2.	Who is responsible for electrical regulations internationally?
3.	At what stage of construction would you normally carry out the second fix of an electrical installation?
4.	In which appendix of the IET Wiring Regulations will you find descriptions of various installation methods?
5.	What is the name of the UK committee that is responsible for the IET Wiring Regulations?
6.	What is the term used for cutting out a channel in a wall so that cable can be installed and later covered (usually by plaster)?
7.	What statutory regulation relates to the need for the electrical supply to be safe and consistent throughout the UK?
8.	Where can you find descriptions and definitions in the IET Wiring Regulations?
9.	What does MHOR stand for?
10.	What set of regulations relate to employees using electrical equipment at work?

Health and Safety 2

After this section, you will have awareness of the following topics:

- ▶ **Risk assessment process**
- ▶ **Safe isolation procedure**
- ▶ **Tool safety**

2.1 Introduction

There is a risk to the health and safety of anybody anywhere. The Health and Safety at Work Act 1974 (also known as HSWA, the HSW Act, the 1974 Act or HASAWA) is a very important piece of legislation that puts the responsibility on employers to provide, so far as is reasonably practicable, a safe working environment for employees and sufficient training so that they can work safely.

This subject always seems to be the least favoured, along with regulatory requirements. However, the implementation of health and safety in the construction industry has significantly reduced the number of accidents, injuries and fatalities over the years.

Health and safety begins at the design stage and continues right through to the end of any project. A great number of factors must be considered, ranging from the location of the construction project through to the environmental impact of the work being carried out. Individual work activities that will take place must also be considered, no matter how small. If there is any demolition work to be carried out, for example, the existing building fabric must be assessed for any hazardous materials, such as asbestos.

2.1.1 What happens when a risk assessment is not carried out?

Example

When I was working as an electrician on a large project in Cambridge, a team installing plasterboard were preparing for the following week's work by loading all their material and tools into the building. This was done on a Friday afternoon. They had decided to stack all the plasterboard sheets into one room on the 4th floor central to the area they would be working in. This meant that there was around two tonnes of plasterboard in a single room in the building in addition to tools and equipment.

Over the weekend a security guard was required to inspect the entire construction site periodically to ensure that there were no trespassers or hazardous situations. While doing his last check at around 05:30 on Monday morning he heard a loud crashing sound close to his location in the building. When he went to inspect he found himself looking down from the 4th floor to the underground car park 20 metres below. The weight of the plasterboard sheets had been more than the floor was designed to withstand and it had collapsed under the weight, taking out the floor in every room below until it hit the concrete floor at the bottom. Not more than two hours later up to 10 construction workers were due to be working in the floors below. This could have had fatal consequences if the floor had held out for a little while longer.

If a risk assessment had been carried out it is likely that the weight of the material would have been taken into consideration and the accident would not have happened, preventing the damage caused and the risk of possible injury.

Consequences of not carrying out a risk assessment

Organisations have a legal duty to put in place suitable arrangements to manage health and safety. If an individual or an employer does not carry out a risk assessment (documented if employing more than five people) there may be legal consequences if an injury occurs as a result of a preventable risk. A risk assessment may have identified the possible risk and injury could have been avoided.

2.2 Risk assessments and method statement

In this section we will explain what a risk assessment is and list the fundamental requirements that an electrician should consider when designing an electrical installation, testing an electrical installation, installing equipment etc.

2.2.1 What is a risk assessment?

Most of us don't realise that we carry out risk assessments every minute of every day. While we are assessing these risks we are immediately thinking of the various steps to take to avoid them. For instance, if we approach a banana skin on the pavement, we would automatically avoid stepping on it as we know we could slip on it and injure ourselves; we have, therefore, carried out a risk assessment and implemented a strategy to lower the risk of injury.

▼ **Table 2.1** An everyday risk assessment: crossing the road

Risk	Risk reduction
Possible collision with a car.	Look both ways and listen.
May trip while standing down from the kerb.	Check the kerb height before stepping down.
Restricted access on the other side.	Plan where you want to cross and check the other side.

In Table 2.1 are all risk factors that we consider could lead to an accident or injury – even though we don't consciously decide to conduct a risk assessment every time we cross the road.

A risk assessment will, in most cases, need to be documented and accompanied by a method statement. If you are working on your own or have less than five employees it is not a legal requirement to produce a written risk assessment. However, this does not mean that a risk assessment should not be carried out.

2.2.2 What is a method statement?

A method statement is a document that details how a task is to be completed and should outline the potential risks that may be involved and include guidance on how to do the job safely. The method statement must also include any measures that have been introduced to ensure the safety of anyone who is affected by the work being done.

> **Example**
>
> You have been asked to install additional lighting in a school classroom. One of the first things that you must complete as an installer/designer is a risk assessment.
>
> When carrying out a risk assessment for the installation of basic electrical equipment the installer will consider where and how the equipment needs to be installed:
> - Is it easily accessible?
> - Will a ladder or tower be required?
> - How heavy is the equipment that must be installed?
> - Will more than one person or lifting device be needed to install the equipment?
> - To what support is the equipment being fixed?
> - Is the building fabric hazardous in any way (for example, is asbestos present)?
> - What is the building used for?
> - Will there be members of the public or employees that will be affected by the work to be carried out?
>
> The more concise a risk assessment, the easier it is to reduce the potential injury or inconvenience that may occur.

We can now look at the various measures that can be implemented to reduce the risks that have been identified. When documenting these measures, you are producing a method statement.

Figure 2.1 is an example of a risk assessment relating to two of the possible risks that may be identified prior to carrying out the small job.

▼ **Figure 2.1** Example of a risk assessment

Risk Assessment

Name: S. G. Sparky

Date: Don't do tomorrow what can be done today   **Time:** None like the present

Location: e.g. college room 7671
MFH Stevenage

Type of installation: e.g. domestic lighting
Commercial

Identify possible risks: e.g. trip hazards, poor lighting etc.	Who will be affected: e.g. staff, general public etc.	Possible measures to reduce identified risks: e.g. clear area, provide additional lighting etc.
1. Is it easily accessible? No! It is on the ceiling around 3 m high. Risk of falling and overstretching.	Installer/s	Provide suitable access equipment e.g. ladder or tower. Provide ladder and tower training for installer. Have an additional person to provide support when installing.
2. How heavy is the equipment to be installed? The fitting to be installed is heavy and will be difficult to lift and potentially cause injury to persons lifting the equipment.	Installer/s	At least two people required. Tower with platform required, installers to be trained in manual handling techniques and suitable PPE provided.

2.2.3 Acceptable risk

If the measures to reduce some of the possible risks completely are expensive and time consuming it may be necessary to categorise this as an acceptable risk and take reasonable measures to prevent the possibility of accident or injury. You can use a risk rating system like the one below, which will help you to help determine the level of severity of the identified risks.

▼ **Table 2.2** Acceptable risk

Risk rating			Risk Action Bands	
Chances of occurring	X Level of injury	= Risk assessed as	Measures of control.	
1. Very unlikely	1. Minor injury	Minimal 1 or 2	Maintain existing measures.	
2. Not likely	2. Slight injury	Low 3 or 4	Review measures.	
3. Likely	3. Serious injury	Medium 6 or 8	Improve measures.	
4. Very likely	4. Major injury	High 9, 12 or 16	Improve measures immediately. May require work to discontinue until additional measures have been applied to reduce the risk.	

Minimal risk: k-rating 1 or 2

If there are measures in place to maintain a minimum risk there is no immediate action required. The risk assessment will be continuously reviewed so that any hazards arising will be identified and rectified.

Low risk: 3 or 4

Even a low risk can be reduced in most situations. There may still be something you can do to reduce the rating to minimal.

Medium risk: 6 to 8

A rating of 6 or 8 is quite risky. You should review existing safety measures and apply as many measures as possible to reduce the risk rating to minimal or at least low.

High risk: 9 to 16

This is obviously quite a risky situation. If you cannot identify sufficient measures to reduce the risk further then it may be necessary to consult with a professional advisor so that adequate provisions can be applied to reduce the risk factor. It may be that additional training or specialist equipment is needed to achieve this.

2.2.4 What next?

The risk assessment doesn't stop there. It is a requirement that the risk assessment will be continuously reviewed for the duration of the work being done. This is necessary as nothing stays the same for long. On a building site there are changes and developments occurring daily, if not hourly. Some risks may be reduced while others are increased. It's important to keep the risk assessment up to date.

> **Remember...**
>
> **A summary of the process**
>
> (a) Assess the possible risks – it is advisable to make a record of these.
> (b) Determine the possible consequences of the risks identified – again, details of the possible consequences should be documented.
> (c) Document who or what will be affected.
> (d) Put together a list of measures that can be applied to reduce the impact or prevent the risk altogether.
> (e) Evaluate the risk assessments – and add to it where necessary e.g. when materials have been moved from one location to another there may be additional risk at the new location.
> (f) Continuously review the risk assessment and update accordingly.
>
> Following these steps should satisfy most organisations that a sufficient risk assessment has been carried out. For apprentices building a portfolio for a vocational qualification it is important to document that these steps have been followed.

▼ **Figure 2.2** Risk assessment flow chart

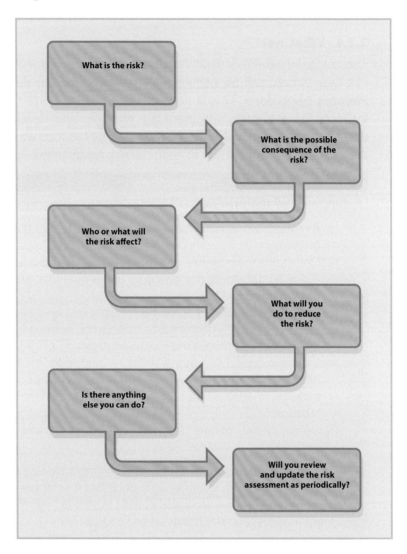

Using the template below, write a risk assessment for a work activity you will be doing in the workshop i.e. installing a lighting circuit, cutting conduit etc.

Remember, you can't remove a risk completely but there are ways to reduce the risk.

Test your knowledge

Risk Assessment		
Name:		
Date:		**Time:**
Location: *e.g. college room 7671*		**Type of installation:** *e.g. domestic lighting*
Identify possible risks: *e.g. trip hazards, poor lighting etc.*	**Who will be affected:** *e.g. staff, general public etc.*	**Possible measures to reduce identified risks:** *e.g. clear area, provide additional lighting etc.*
1.		
2.		
3.		
4.		
5.		

2.3 Safe isolation procedure

The safe isolation procedure is a process that, when carried out correctly, will reduce the risk of electric shock, burns or death of any person likely to come into direct contact with an electrical installation, whether that installation is in the process of being installed or is under maintenance. It is also very important to inform the person responsible for the premises of the implications.

If the procedure has not been carried out correctly, for example, and the protective device has not been locked off and labelled with an approved device, there is the possibility that somebody may switch the supply back on while an electrician is working on the circuit. This could have fatal consequences.

All persons working in the electrical industry are required to be fully competent in carrying out this procedure and students can expect to be asked to demonstrate their ability to do so during practical assignments and assessments. By using this Guide you will understand how to meet the current requirements of awarding bodies.

2.3.1 What is Guidance Note GS38?

Guidance Note GS38, published by the HSE, is the recommended requirements for test leads and instruments that are intended to reduce the risk of electrocution, burns or death when carrying out tests on electrical systems.

In some circumstances, this may seem a little over the top, but why risk the possibility that you will potentially suffer a fatal electric shock when you can use test equipment that has safeguards in place to prevent it?

As well as ensuring that the equipment is safe to use, the person carrying out the test must be adequately skilled to use the equipment too.

▼ **Figure 2.3** Test leads

▼ **Figure 2.4** Safe isolation procedure

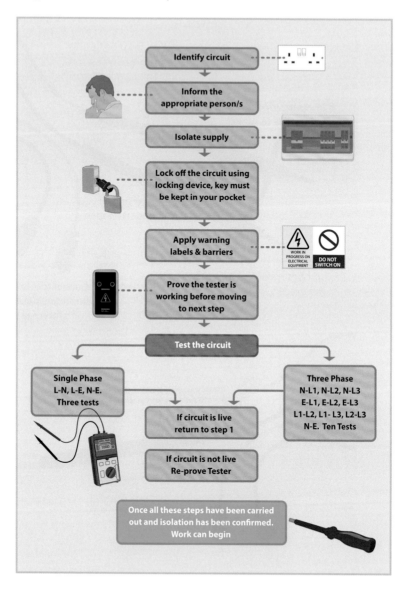

2.4 Tool safety

A basic toolkit for an electrician will normally consist of:

(a) a set of insulated (electrical) screwdrivers;
(b) a torque screwdriver;
(c) a pair of side cutters;
(d) a pair of pliers;
(e) a pair of wire strippers;
(f) a hacksaw;
(g) a measuring tape;
(h) a set square;
(i) a ruler;
(j) a file;
(k) a pair of grips; and
(l) a hammer.

All tools have the potential to cause injury if not used correctly. Guidance on how to use the tools is provided by the manufacturer. Training for the use of tools should be provided by instructors and/or employers.

▼ **Table 2.3** Various hand tools used by electricians (Images courtesy of Stanley Tools)

Grips	Grips (also known as 'pipe grips') are a fantastic tool that, when used correctly, are designed to tighten their grip on whatever it is they are turning. It is important to understand that they are designed to turn in one particular direction, which causes them to increase the pressure on what they are holding. If used incorrectly they will take more effort and will likely slip and potentially injure you.
Hacksaw	A hacksaw or junior hacksaw is a fairly simple but valuable tool.
	BS 7398:1991 identifies the requirements for hand hacksaw frames, BS 1919-1:1993 identifies the requirements for hacksaw blades and BS 6271:1982 identifies the requirements for miniature hacksaw blades.
	Maintaining a hacksaw is also fairly simple. Make sure it's stored safely and, when the blade gets blunt, replace it. Also remember to lubricate the tightening mechanism. When replacing the blade ensure that it's fitted the correct way around (if you look closely at the blade you will see that the points all face one direction; make sure that they are pointing away from the handle when you fit the blade).
	Larger hacksaws usually have adjustable tension. This must be set to a suitable level for the tool to work efficiently.
Hammer	A hammer can be used in a number of situations. Of course, its true purpose is to drive nails into wood or similar material. However, it is known to be used for making minor adjustments to the position of equipment. This is not, however, recommended!
	BS 876:1995 identifies the requirements for hammers.

Measuring tape 	Not the most dangerous tool in the box but, nevertheless, it needs to be used responsibly. BS 4035:1966 identifies the requirements for linear measuring instruments for use on building and civil engineering constructional works. This Standard encompasses steel measuring tapes, steel bands and retractable steel pocket rules. When taking long measurements the tape can retract quickly and potentially cause cuts if it comes into contact with flesh. Tape measures are susceptible to damage when they are extended beyond the maximum measurement they are designed to take. When this happens it can be very difficult if not impossible to repair. One key point to remember – measure it twice, cut it once!
Pliers and side cutters 	Side cutters and pliers are probably the most commonly used tools in an electrician's toolbox. As well as cutting a variety of sizes of cable they are often used to strip off cable sheathing. Pliers can also be used to hold nuts and bolts in place to make it easier to get the thread started. BS 3087-10:1991 sets out the requirements for pliers and side cutters (or 'snips', as they are sometimes called). Just as with screwdrivers, it is important to make sure that the insulation of your side cutters and pliers is in good condition. If there's a breakdown of the insulation on the handle of the tool and it unintentionally comes into contact with a live part it could present a risk of electric shock to the user.

Screwdrivers/ torque screwdrivers	Every electrician will need to have a set of insulated screwdrivers. BS 2559:1994 identifies the requirements for insulated screwdrivers – yes, there is a standard for screwdrivers!
	This is necessary for safety – if, for example, you were to purchase a screwdriver that did not comply with the Standard, it may not have the required insulating properties and could present a risk of electric shock to the user. Manufacturers of consumer units and similar switch-gear assemblies may require a specific torque setting to be used when terminating conductors.
	To maintain your insulated screwdrivers, it is advisable that you have a suitable means of storing them separately to the rest of your tools. This will avoid them being damaged and the possibility of the insulation failing. If you have a torque screwdriver it is important to have the settings calibrated in accordance with the manufacturer's instructions.
	If the insulation on your insulated screwdriver is damaged, which can happen when it is used for various other activities (such as removing knock-outs from consumer units or checking for a fixing in ceilings by pushing it through the plasterboard), you must consider replacing the screwdriver as it may no longer be sufficient to protect you from the risk of electric shock.
Stocks and dies	Stocks and dies are used to thread conduit so that it can be attached by a screwing method to other equipment, such as other pieces of conduit, switch and socket-outlet boxes etc.
	BS 1813:1974 identifies the requirements for stocks and dies, including the dimensions of conduit dies, die stocks and guides.
Wire strippers	Wire strippers are used to remove the insulation from cable. They can be adjusted to be used on various sizes of cable. Wire strippers can be easily damaged and should be kept separate from heavy tools such as grips, hammers etc.

▼ **Table 2.4** Various power tools used by electricians (Images courtesy of DeWalt)

110 V drills	110 V is commonly used on constructions sites, although electricians who work in a variety of environments may have a portable 230/110 V transformer so that they can use 110 V in a variety of places.
230 V drills	Mains voltage drills can be used in most environments by plugging into a 13 amp socket-outlet.
Battery drill	Battery drills are very common in construction. Technology has improved battery life and reduced the weight of such drills significantly. Battery drills can be used for drilling small holes for fixings and access. Some have settings so that they can be used to chase walls (see 'chasing' in Table 1.6). BS EN 50260-2-1:2002 identifies the requirements for the safety of hand-held battery-powered motor-operated tools and battery packs, and particularly the requirements for drills.
Chop saw	These are used to cut conduit, slotted, channel, cable tray, trunking and much more. They are quite heavy-duty cutting tools that go easily through light metal.

Jigsaw	A jigsaw is used to cut through various types of material including metal and is normally used on a flat surface. Electricians might use this type of saw to cut out entry holes in large distribution boards or lengths of trunking.
Threading machine	When installing conduit, the ends must be screwed together and to the equipment. This is normally done by hand with stocks and dies.

2.4.1 Choosing the correct tool for the job

Tools are designed to be used for specific tasks and with specific pieces of equipment. Many manufacturers of tools and components design their products to be used only with their own equipment. This helps to reduce the damage that may be caused by the incorrect use of tools and to maintain a good standard of installation.

Various consequences may result from using the incorrect tools:

(a) Damage to terminal screws

Using the wrong type of screwdriver can cause damage to terminal screws as they won't fit correctly. It's important to feel if the screwdriver fits the screw snugly and it is therefore important to know the differences between slotted, PZ, PH, etc. Any excess movement will be liable to cause damage to the screw head as it is tightened.

(b) Undertightening or overtightening terminal screws

If an undersized screwdriver is used it may be difficult to apply the required torque to ensure a good termination. If an oversized screwdriver is used it may overtighten the screw and cause damage to the terminal that may cause the conductor to become loose.

(c) Damage to the appearance of accessories and equipment

Equipment is expensive and many electrical items are chosen for their appearance. If the wrong type of tool is used to install the equipment, it is more likely to damage the equipment.

(d) Loose glanding of cables at entry points

When connecting cable glands it can sometimes be difficult to reach with ordinary tools and specialist equipment must therefore be used. Some glands form part of an earthing system as well as forming an integral part of a cpc. Not terminating them properly can affect the continuity readings when it comes to testing and fail to provide a sufficient electrical connection to earth.

(e) Injury to the user

Tools can be dangerous, especially when they aren't used for their intended purpose. For example, using a drill bit instead of a de-burring tool to remove excess material could result in the drill bit breaking, becoming a projectile and causing injury.

(f) Damage to the tools

Tools are expensive. If they are not being used for their intended purpose they are more susceptible to damage and replacement. Using a large screwdriver as a chisel would certainly reduce the life of the screwdriver.

	Test your knowledge
1.	List five tools and the potential risk of injury that may occur when using them.
2.	The insulation on tools designed for electrical work is to protect the user from what?
3.	When safely isolating a circuit, where should the key be kept while work is being carried out?
4.	True or False: Once a risk assessment has been carried out and documented it does not need to be looked at until work has been completed.
5.	What tool should be used to remove insulation from a cable so that it can be terminated correctly?

Generation and Transmission 3

After reading this section, you will have awareness of the following topics:

- ▶ What is required to generate electricity?
- ▶ The characteristics of the electricity generated in the UK
- ▶ How electricity is transmitted
- ▶ How electricity is distributed
- ▶ How electricity is measured

Although the generation, transmission and distribution of electricity are not covered by the IET Wiring Regulations, it is essential to understand these processes in order to appreciate their relevance to electrical installations in general.

3.1 What is used to generate electricity?

Electricity is generated in the UK using various different methods. The most common of these are listed below:

Fuel	Type	Environmental impact
Coal	Fossil fuel – non-renewable	High CO_2 emissions
Gas	Fossil fuel – non-renewable	High CO_2 emissions
Oil	Fossil fuel – non-renewable	High CO_2 emissions
Nuclear	Fission – non-renewable	Low CO_2 emissions
Solar	Renewable	Low CO_2 emissions
Wind	Renewable	Low CO_2 emissions
Hydro	Renewable	Low CO_2 emissions
Wave	Renewable	Low CO_2 emissions

Fuel	Type	Environmental impact
Petrol/diesel generator	Fossil fuel – non-renewable	High CO_2 emissions

Note: The term 'carbon neutral' implies that the CO_2 that is being produced has been balanced out in some way, for example, by the planting of trees that absorb CO_2 and turn it back into oxygen.

3.1.1 Power stations

The basic principles of generating electricity at any power station are mostly the same. Some form of fuel, such as coal or gas, is burned in a furnace to heat water into steam in a boiler. The steam travels through pipes at high pressure and is used to turn a turbine at high speed which turns a generator. Steam is used to power a steam train in the same way, but instead of powering wheels on a track steam in a power station turns a generator. Generators are made up of coils of copper wire and iron to form electromagnets. When these electromagnets are turned within one another it creates electricity. The electricity generation process is outlined in Figure 3.1.

▼ **Figure 3.1** Generation of electricity

Electricity in the UK must be supplied at a frequency of 50 Hertz in accordance with ESQCR

3.1.2 Hydro and wind generation

Hydro and wind generation processes use generators that work on the same principle as the ones in power stations except that they are powered by the flow of wind and water instead of pressurised steam.

3.1.3 Solar power

Solar power, on the other hand, works quite differently. A variety of materials are combined using specialised techniques to produce what are known as solar panels. These solar panels have the ability to use energy from the sun to create electrical current. By controlling this current it is possible to harness the energy from the sun and generate electricity.

3.2 What are the characteristics of the electricity generated in the UK?

The ESQCR require that the suppliers of electricity in the UK provide users with a standard power supply of similar characteristics across the whole of the UK. There are a number of reasons for this. One good example is electric clocks: in order for a clock to maintain accurate time the power supply must be at a certain frequency. In the UK that frequency is 50 Hz, which means that the alternating current travels back and forth 50 times a second and the low voltage supply shall be 230 V between line and neutral. For three-phase supplies it will be 400 V between live conductors and 230 V between line and neutral.

The declared nominal supply voltages may vary within permitted tolerances of +10 % and -6 %. 230 V became the UK nominal voltages for single phase supplies and 400 V for three phase in 1995.

3.3 How is electricity transmitted?

3.3.1 The journey of electricity

Electricity is produced by generators at power stations at about 25,000 V. This is not enough to send it long distances along transmission lines, so the electricity needs to be changed through a step-up transformer at the power station that increases the voltage to between 275,000 V (275 kV) and 400,000 V (400 kV) see Figure 3.2. Electricity travels long distances more efficiently at higher voltages.

▼ **Figure 3.2** Transmission from power station to National Grid

When the electricity leaves the transformer it travels through the National Grid (sometimes just called the 'Grid'). The Grid is the network of conductors connected to one another across the UK that transmits electricity from power stations to towns and cities. It is designed to be interconnected with every power station in the UK so that if, for some reason, one power station fails, the towns and cities in that area will still have power. The wires that carry the electricity in the Grid are called transmission lines. These lines are supported across the country by tall electricity transmission towers (see Figure 3.3).

▼ **Figure 3.3** Examples of transmission towers

3.3.2 Transmission

Before electricity from the Grid can be used by factories, shops and homes it must be transmitted to substations so that the electricity can once again be transformed. This time, however, it's the other way around as the voltage is much too powerful to use in our homes and businesses and so must be reduced.

Transformers in the substations reduce the very high 275 kV or 400 kV to 132 kV before it enters the regional distribution network. Regional distribution networks carry electricity to substations; the voltage is again reduced from 132 kV to 11 kV. The 11 kV network supplies towns, industrial estates and villages, as well as some industrial premises that have large electricity requirements, such as car manufacturing plants.

3.3.3 Distribution

▼ **Figure 3.4** Examples of distribution transformers

The final stage is the local substations and distribution networks. The grey boxes that can be seen in fenced-off areas with high voltage signs posted, are the local substations and distribution transformers.

At this stage the voltage is reduced to 400 V or 230 V. Many commercial outlets, such as shops and supermarkets, will have a 400 V supply, while domestic premises (homes) will normally have 230 V supply. If you look at the technical information written on your household appliances, such as the kettle and TV, you will see the voltage will be 230 V.

3.4 How do we measure what we use?

▼ **Figure 3.5** Electricity meters

Unfortunately, electricity is not free. In fact, it is quite expensive with the average household spending around £1,000 a year on their electricity bill.

So how is it measured? You will be familiar with your consumer unit at home. Next to it you will generally find some other equipment and thick cable. The one to look out for is either going to have a spinning disc behind a glass screen or a little flashing light with an LCD display with lots of numbers on it. This is your electricity meter. The meter records the power used in kilowatt hours (kWh). The electricity supplier will sell the power to the customer for an agreed price. The details of the meter need to be provided to the electricity supplier so that they can accurately charge the customer for what they use.

Whenever electricity is used in a property, it will pass through these meters, recording every little bit of power that is being used. Remember that when you leave the bathroom light on next time!

Now let's look at the whole picture (Figure 3.6).

▼ **Figure 3.6** Transmission and distribution network

	Test your knowledge
1.	Name two types of fuel used by power stations that will run out if we continue using them.
2.	What does ESQCR stand for?
3.	Why is voltage increased when electricity is being transmitted?
4.	Where might you find a distribution transformer?
5.	What is the frequency that electricity is generated at in the UK?
6.	When electricity is measured at the consumer's installation, what is the unit of measurement used?
7.	What is the greatest benefit of the National Grid?
8.	What does a step-up transformer do?
9.	What does a step-down transformer do?
10.	What voltage would you usually find in a domestic property in the UK?

Supply 4

After this section, you will have awareness of the following topics:

- ▶ Supply intake arrangements
- ▶ Live conductors
- ▶ Current
- ▶ Earthing arrangements
- ▶ Tails
- ▶ Consumer unit
- ▶ Ways and modules

4.1 Supply intake arrangements

Before any work is carried out within an electrical installation, details relating to the supply must be looked at and recorded, because such details can change the way in which work is done on the installation and what type of protection is necessary. The supply details can be looked at in three ways:

(a) live conductors – how many are there?
(b) current – what is the maximum rating?
(c) earthing arrangement – what type of earthing system is used?

4.2 Live conductors

Supplies to smaller domestic and commercial installations is usually 230 V single-phase, meaning that there are two live conductors (line and neutral). Always remember that a neutral carries current, when the circuit is in use, so it is classed as a live conductor.

Larger installations, which normally need more than 100 A, will usually be provided as three-phase supply. This allows the larger overall load to be spread over the three-phase supply within the installation, thereby reducing the risk of overloading one phase.

▼ **Table 4.1** Live conductor arrangements within the UK

Type	Arrangement	Location
Single-phase 230 V between Line and Neutral	Two-wire. An earthing conductor is likely to be present but it's not deemed to be a live conductor.	Dwellings, small shops or small commercial installations.
Three-phase 400 V between Line and Line	Three-wire (no neutral).	Installations are rarely three-phase only. Often a neutral conductor is required as single-phase (i.e. line-neutral) circuits will be needed for lighting and appliances.
Three-phase & neutral 230 V between Line and Neutral : 400 V between Line and Line	Four-wire.	Larger retail outlets, larger commercial or industrial buildings.

4.3 Current

Another factor that affects the supply is the current rating, which is often referred to as 'maximum demand'. Most average-sized houses in the UK will have a maximum demand of less than 100 A and many houses in a road or estate will be supplied from the same single sub-station transformer. Most commonly, house supplies are rated at 80 A.

Larger installations that have a maximum demand over 100 A will likely be supplied using three-phase supply in order to spread the load in the supply. Where a building needs a supply over 400 A per phase, it may have its own sub-station transformer installed on site.

4.3.1 The Distribution Network Operator

The maximum demand for an electrical installation will be governed by the rating of the Distribution Network Operator's (DNO) service cut-out fuse. This means that, where an existing installation is to be added to, you must always check that the supply has enough capacity for the extra load.

Where a new building is to be given a supply, the DNO needs to know the proposed maximum demand for the electrical installation so that they can provide an adequate supply. It would be the job of the designer to provide this.

Typical current ratings of DNO service fuses are 63 A, 80 A, 100 A, 125 A, 150 A, 200 A, 300 A, 400 A and 500 A.

The DNO's service fuse will be fitted inside the electrical service head (cut-out), which is the enclosure into which the supply cable is terminated before the meter. Most DNO fuses will be in a fuse carrier that is 'sealed' to the service head, meaning it can only be removed or replaced by a person authorised by the DNO. The seal is a small wire which, with a special sealing crimp tool, locks the fuse to the service cut-out, so it has to be broken if the fuse is removed. Connections to meters will also be sealed in this way. If you need to remove a fuse from a service head, you must always contact the DNO.

It is important to know the supply rating, but it is not always easy to see what the DNO fuse rating is. In this situation you should always contact the DNO, who can tell you all the supply details, such as the rating. The DNO has a statutory obligation to provide a designer with details of the supply, including the earthing arrangements.

4.4 Earthing arrangement

The earthing arrangement reflects how the electrical installation and supply is earthed. In the UK there are three common earthing arrangements, which are described using abbreviations (see

Sections 4.4.1-4.4.3). The type of earthing arrangement given by the DNO depends a lot on where the installation is located and/or how old the supply cable is.

There are three common arrangements.

4.4.1 TN-C-S arrangement

The DNO earth and neutral are combined in the same conductor of the supply. In this arrangement, as the supply neutral is also the earth, it is called a PEN (Protective Earthed Neutral) conductor. Once the electrical supply is terminated into the service head, the PEN conductor is split into two terminals, one is neutral, the other is earth.

This is the most common earthing arrangement in the UK. It is also called PME (Protective Multiple Earth), because it could have more than one 'source' of earth electrode along the length of the buried supply cable in the street. For this system, the DNO will declare an expected maximum external earth fault loop impedance (Z_e) of 0.35 Ω.

This system may be identified by looking at the DNO service head; if the earthing conductor comes from inside the service head, it is likely to be PME. (If in doubt, the DNO should be contacted for clarification.)

▼ **Figure 4.1a** TN-C-S arrangement

4.4.2 TN-S arrangement

The DNO earth and neutral are separate conductors in the supply and installation. The supply earth is normally the metallic sheath or armour of the supply cable.

This arrangement is common in older houses. For this system, the DNO will, in most circumstances, declare an expected maximum external earth fault loop impedance (Z_e) of 0.8 Ω.

This system may be identified by looking at the supply cable below the service head − the earthing conductor is often clamped to the metallic sheath or armour of the supply cable. (If in doubt, the DNO should be contacted for clarification.)

▼ **Figure 4.1b** TN-S arrangement

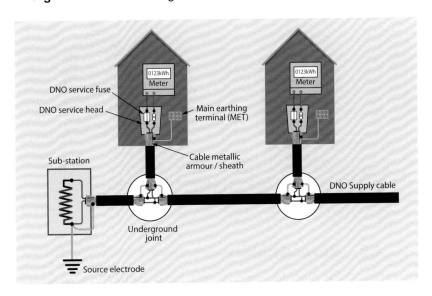

4.4.3 TT arrangement

In some cases, the DNO will not provide an earth and the responsibility lies with the consumer to provide this in the form of an earth electrode. The two-wire supply to this type of installation will normally be carried via overhead cables.

This arrangement is more common in rural areas where transformers are mounted on poles serving, for example, one or two dwellings. More and more of these are being converted by the DNO to TN-C-S arrangements.

This system is identified by the earthing conductor being connected to an earth electrode in the ground. External earth fault loop impedances can vary greatly depending on many factors but values of less than 200 Ω are regarded as stable values of Z_e. As the ground is the conductor between earth electrodes, problems can arise when the ground freezes in winter or dries out in summer, so keeping Z_e at less than 200 Ω improves stability.

▼ **Figure 4.1c** TT arrangement

4.5 Responsibility

Once the electrical supply has been metered, it generally stops being the responsibility of the DNO or the supplier and becomes the responsibility of the consumer. The consumer is the person who owns the electrical system on the property – in simple terms, the person who owns the electrical installation. It is important to understand that the supplier and the DNO are not the same.

Depending on where the meter is located, what happens next can vary. In some situations, more likely newer buildings, there may be an isolation switch between the meter and the consumer unit. This isolation switch may have been fitted by the DNO or by the electrician who carried out the installation work.

If you are fitting a new installation, it is always a good idea to fit an isolation switch between the meter and the consumer unit so that work can be carried out on all parts of an installation, including inside the consumer unit with the supply completely isolated.

Isolation switches are also fitted with internal fuses (switch-fuses) where the consumer unit is located some distance away from the meter point.

▼ **Figure 4.2** Typical intake layouts and responsibilities

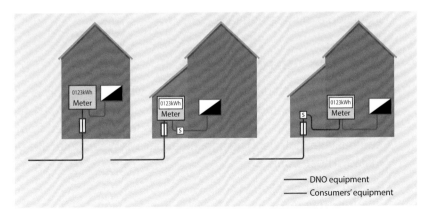

4.6 Tails

Tails is the name given to the single-core, insulated and sheathed cables that go into and out of the meter. Two terms are used:

(a) meter tails: connecting the supplier's cut-out to the meter. These are usually owned by the meter operating company.
(b) consumer's tails: connecting the meter to the consumer unit or distribution board. These are the responsibility of the installation owner.

Single-core cable is used to give flexibility and allow easy installation. The cable has insulation (the brown or blue covering of the cable), which provides basic protection against electric shock. The cable must also have an outer sheath, which provides mechanical protection against physical damage (this is usually the outer grey covering of the cable).

Tails should be as short as possible to avoid damage and should be clipped or installed in trunking to avoid them being moved. When a cable is moved this can put stress on the termination and, over time, cause the termination to deteriorate, i.e. the cable can become loose in the terminal. The sheathing must cover the entire cable, especially when inside a metal consumer unit, right up to the point where they connect to the main switch inside the consumer unit.

If the consumer unit is located away from the meter point, a much better cable, such as steel-wire armoured, should be used instead of insulated and sheathed tails.

4.7 The consumer unit

Generally, the first unit in an installation, where the supply is split into circuits, is called the consumer unit. Regulation 421.1.201 requires all consumer units in domestic dwellings to be fabricated from a non combustible material. The consumer unit will have protective devices inside. These could be:

   (a) circuit-breakers; or
   (b) circuit-breakers and RCCBs; or
   (c) residual current breakers with overload protection (RCBO); or
   (d) fuses.

These protective devices protect the circuits. The consumer unit will also have a main switch that can be used to isolate the entire installation. In a domestic dwelling, the main switch must be double-pole, meaning that it isolates both line and neutral. In some three-phase installations that are part of a TN earthing arrangement, the main switch may be three-pole, meaning that it switches the three line conductors only. In this situation, the neutral can be isolated by a link in the consumer unit.

Three-phase installations having a TT earthing arrangement must have four-pole switches isolating all live conductors.

In a domestic dwelling, the main switch in a consumer unit must be the type that is rated to switch the full load current. Some isolator devices that act as main switches are not rated for full load current and, if they are used regularly to switch full load, they will eventually fail.

You may hear a consumer unit called many different things such as 'fuse board', 'distribution board' (DB) or even 'fusebox'. A DB is often the name given to further units on the end of a distribution circuit. A distribution circuit is a circuit that supplies a distribution board, whereas a final circuit is one that supplies current using equipment such as lights and heaters or socket-outlets.

▼ **Figure 4.3** Distribution and final circuits

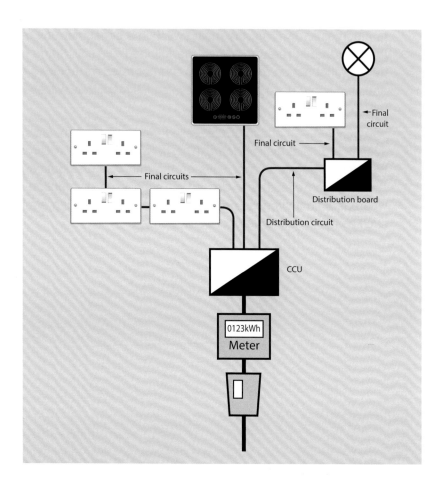

4.7.1 Ways and modules

When you buy a consumer unit, the amount of circuit-breakers or main switches that can be installed are described differently.

(a) **Modules:** This is the number of components that can be fitted into the consumer unit. For example, a main switch controls both line and neutral so is classed as a two-module device. A circuit-breaker is usually single-pole, so it would be one module. If you needed a consumer unit to control six circuits and have a main switch, you would need an eight-module board.

(b) Ways: This describes the number of circuits that can be controlled. A six-way single-phase board could house six circuit-breakers as well as the main switch.

A six-way three-phase board can house six three-phase circuit-breakers or eighteen single-phase circuit-breakers. It could also house a combination of the two, such as two three-phase (six ways) and twelve single-phase (twelve ways).

▼ **Figure 4.4** Six-way consumer unit with eight modules

Split-way

Some consumer units are known as split-way boards. This means that some of the modules can be protected by one RCD and the others by another RCD. Split-way boards should always have one main switch that can isolate all circuits.

▼ **Figure 4.5** Split-way board with three ways protected by one RCD and three ways protected by another (12 modules in total)

	Test your knowledge
1.	What two types of connection are described for a three-phase 3-wire connection according to Part 3 of the IET Wiring Regulations?
2.	What does PEN stand for with regard to TN-C-S earthing arrangements?
3.	List the information that a supplier must provide on request in relation to the characteristics of a supply in accordance with Section 313 of the IET Wiring Regulations.
4.	What type of cable should be used for 'tails' in a domestic property where the cables are installed in free air with no further containment system?
5.	Which regulation of the IET Wiring Regulations states that a consumer unit within a domestic (household) installation should have an enclosure made from non-combustible material?
6.	List three reasons given in Part 3 of the IET Wiring Regulations for dividing an installation into circuits.
7.	Six circuits are to be supplied from a consumer unit that incorporates a double-pole main switch. What is the minimum number of modules required in the consumer unit?
8.	Which regulation from Part 5 of the IET Wiring Regulations states that devices installed within an assembly must have their compatibility verified?
9.	What is the maximum rating of a socket-outlet that can be used as a functional switching device according to Table 537.4 of the IET Wiring Regulations?

Protection and Isolation 5

After this section, you will have awareness of the following topics:

- ▶ Types of protective devices
- ▶ Uses of protective devices
- ▶ Selectivity
- ▶ Isolation and switching
- ▶ Devices used for isolation and switching
- ▶ Reasons for isolation and switching

5.1 Protective devices

A protective device can provide protection against the various types of overcurrent, such as overload, short-circuit and earth fault. An *overcurrent* is any current that exceeds the current-carrying capacity of the circuit conductors. Some protective devices can provide protection against very small earth leakage currents (more commonly referred to as *protective conductor current*). To understand how these devices operate, we first need to define the types of overcurrent.

(a) **Overload:** a current that is more than the normal load current for fairly short durations. There is no fault in the circuit, just too much equipment being used at once.

(b) **Short-circuit:** where a fault happens between live conductors, such as line to neutral or line to line. The fault is assumed to be of negligible or very low impedance, which would develop high fault currents.

(c) **Earth fault:** where a fault happens between line and earth, creating an electric shock risk. The fault is assumed to be of negligible or very low impedance which would develop high fault currents.

(d) **Neutral fault:** where a fault occurs between neutral and earth. These often go undetected but can cause RCDs to operate.

In addition to the above, protective devices also have different ratings. These are:

(a) **Rated short-circuit capacity (breaking capacity) (I_{cn}):** this is the amount of current that a fuse can take before it explodes or seriously damages the carrier and its surroundings. In the case of a circuit-breaker or RCBO, faults higher than their rated capacity could weld the contacts together so that it won't stop the current flowing.

(b) **Nominal rating (I_n):** this is the level of current that the device can handle in continued service, for example, a 32 A circuit-breaker can easily carry 32 A load current for the lifetime as specified by the manufacturer without deterioration or loss of function.

(c) **Current causing effective disconnection (I_a):** this is the value of current that would cause the device to disconnect in the time required, for example, a BS 88-2 fuse with a 6 A rating will disconnect in 0.4 seconds at 18 A. We can determine this using the time current graphs in Appendix 3 of the IET Wiring Regulations or using the manufacturer's data.

▼ **Figure 5.1** Time/current graph as shown in Appendix 3 BS 7671

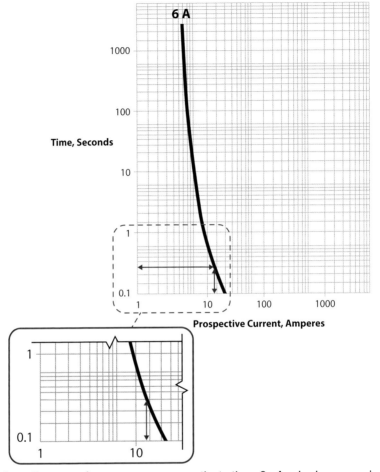

Using the graph, you can see that the 6 A device requires approximately 18 A to disconnect in 0.4 seconds.

You must always remember that a protective device such as a circuit-breaker may have a nominal rating of 16 A, but it will never disconnect at 17 A or 18 A. In the case of a type B 16 A circuit-breaker, an approximate current value of 23 A will probably operate the device and disconnect the circuit in around 3 hours, but it needs 80 A or more to disconnect it quickly.

The following table sets out the common protective devices used in many electrical installations as well as the relevant Standard.

▼ Table 5.1 Protective devices and relevant Standards

Protective device	Standard
High rupturing capacity (HRC) fuses (also called high breaking capacity (HBC) fuses)	BS 88-2
Cartridge fuses	BS 88-3
Semi-enclosed, rewireable fuses	BS 3036
Plug fuses	BS 1362
Circuit-breakers	BS EN 60898
Residual current breaker with overload (RCBO)	BS EN 61009
Residual current device (RCD)	BS EN 61008

Each type of device is different in the way it performs and handles faults or overloads and a designer will select a particular device based on those differences.

5.1.1 High rupturing capacity (HRC) fuses

As the name of these fuses (also known as high breaking capacity (HBC) fuses) suggests, they can handle very high fault currents as they have a high breaking (rupturing) capacity. Inside the porcelain body of the fuse, the fuse element is surrounded by sand. When a large fault current causes the fuse element to blow, the heat from the blast is used to turn the sand into glass as it absorbs most of the energy. Because these fuses can absorb a great deal of energy, they can safely handle fault currents up to 80 kA (80,000 A).

These fuses may be classed as 'System E', meaning they are clipped into place, or 'System G', meaning they do not have the end fixing lugs and are clipped into place.

These fuses come in a wide range of nominal ratings (I_n) from 3 A up to 500 A and even higher.

There are two protection categories of HRC fuse:

(a) gG fuses: these are general use fuses for standard circuit protection. They will operate at lower currents, typically around 7 times their nominal rating.

(b) gM fuses: these are motor rated, which will allow for a surge current that is common to motors starting or discharge lighting starting. These will need higher fault currents to operate.

HRC fuses are normally found in large commercial or industrial buildings.

▼ **Figure 5.2** HRC fuse

5.1.2 Cartridge fuses

These fuses are similar to HRC fuses but are intended for use by ordinary persons, so are more common in domestic dwellings. You may come across them in older consumer units in houses and they may be marked as BS 1361, which was the old British Standard number for them.

They generally have a breaking capacity of 16 kA and the nominal ratings (I_n) range from 5 A to 100 A.

Before circuit-breakers were commonly found in dwellings, these fuses were seen as better alternatives to rewireable fuses, so in older electrical installations you may see these fuses alongside rewireable fuses in the same consumer unit.

▼ **Figure 5.3** Cartridge fuse

5.1.3 Plug-top fuses

You will find these fuses in every 13 A plug-top or fused connection unit. They range from 1 A to 13 A, but the two common ratings are 3 A and 13 A. Their main purpose is to protect an appliance's flexible cable.

▼ **Figure 5.4** Plug-top fuse

5.1.4 Semi-enclosed fuses

Commonly called rewireable fuses, these devices are much less common today but they are still recognised by the IET Wiring Regulations as suitable. They used to be the most common type of fuse in houses and commercial installations but they are being phased out of use in preference to circuit-breakers because:

(a) semi-enclosed fuses have a low breaking capacity – typically 1-1.5 kA;
(b) semi-enclosed fuses may contain asbestos in the fuse carrier;
(c) semi-enclosed fuses are not very user-friendly when the wire needs replacing; and
(d) it is very easy to put the wrong fuse wire rating in, for example, a 30 A wire into a 5 A fuse carrier for a 5 A circuit.

▼ Figure 5.5 BS 3036 Semi enclosed fuse

5.1.5 Circuit-breakers

Circuit-breakers are probably the most common protective devices. They are user friendly and reliable. 'Circuit-breaker' is the term covered by BS EN 60898, however, manufacturers will still refer to them in three general forms:

(a) **circuit-breakers (cb):** these devices to BS EN 60898 come with nominal ratings between 2 A and 125 A and you will see them in all new installations.
(b) **miniature circuit-breakers (mcb):** these devices to BS 3871 have been superseded by BS EN 60898 devices but you will still see them on many older installations. They

come with nominal ratings between 3 A and 80 A and are commonly used in many types of installations as protection for final circuits. The word 'miniature' was dropped when BS EN 60898 devices replaced BS 3871 devices.
(c) **moulded case circuit-breakers (mccb):** these are larger circuit-breakers and are usually used in the large panel boards that protect distribution circuits.

Instead of a breaking capacity, circuit-breakers have two ratings:

(a) I_{CS}: this is the value of fault current that can be safely handled by a circuit-breaker without it being damaged, and with it remaining usable (serviceable) after the fault.

(b) I_{CN}: this is the value of fault current that can be safely handled by a circuit-breaker but it will not remain serviceable and will need replacing. Any fault current above the I_{CN} rating will be dangerous and could cause either the circuit-breaker to explode or the contacts to weld together, meaning that the fault current will not be disconnected.

For example, a circuit-breaker may have an I_{CS} of 6 kA and an I_{CN} of 10 kA. From Figure 5.6 below, we can see that a fault current up to 6 kA will be safely interrupted by the protective device, which remains usable afterwards. Any fault between 6 kA and 10 kA will be safely interrupted by the protective device but will require replacing after the fault as it may be damaged inside. Any faults above 10 kA could cause the protective device to either explode or weld the contacts together.

▼ **Figure 5.6** Relationship between I_{CS} and I_{CN} for circuit-breakers

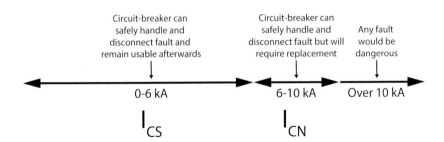

Operation of circuit-breakers

Circuit-breakers operate in two ways:

(a) **via a magnetic trip:** this is a solenoid within the circuit-breaker where a pre-set value of overcurrent will create a magnetic field strong enough to disconnect the device instantly (within 0.1 seconds). The magnetic trip is identified as component 1 in Figure 5.7; and

(b) **via a thermal trip:** this is where an overcurrent heats up a thermal device, such as a bi-metallic strip, causing disconnection. This process takes longer depending on the value of the overcurrent. The thermal trip is identified as component 2 in Figure 5.7.

▼ **Figure 5.7** Internal components within a circuit-breaker

These two ways of tripping a circuit-breaker can be seen by the time current graphs for circuit-breakers in Appendix 3 of the IET Wiring Regulations.

▼ **Figure 5.8** Time current graphs for circuit-breakers in Appendix 3 of the IET Wiring Regulations

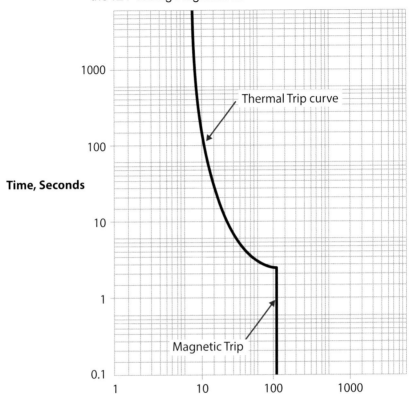

Types of circuit-breaker

There are three types of circuit-breaker. The value of the magnetic trip will vary depending on the type selected. Some loads, such as fluorescent luminaires, draw a surge current when they first start up. This surge current could be seen as a fault by a type B circuit-breaker, so a type C would commonly be used instead as the magnetic trip is set higher meaning more current is required to instantaneously trip the device. Because the thermal trip takes longer to react, the circuit-breaker doesn't trip with the surge current. This surge current should also be taken into consideration when selecting the switches for the circuit so that they can be selected with the appropriate rating.

The three types of circuit-breaker are:

(a) **Type B:** the magnetic trip is set to approximately five times the nominal rating, so a device that has a nominal rating (I_n) of 10 A will trip instantly (within 0.1 seconds) with a fault current exceeding 50 A. This type would typically be used to protect general circuits, such as socket-outlets or fixed heating appliances.

(b) **Type C:** the magnetic trip is set to approximately 10 times the nominal rating, so a device that has a nominal rating (I_n) of 10 A will trip instantly with a fault current exceeding 100 A. These devices would be used to protect loads that surge on start, such as discharge lighting or motors.

(c) **Type D:** the magnetic trip is set to approximately 20 times the nominal rating, so a device that has a nominal rating (I_n) of 10 A will trip instantly with a fault current exceeding 200 A. This type should only be used on very specialist loads, such as welding sets that create high current surges for short durations.

5.1.6 Residual current device (RCD)

RCDs are different from circuit-breakers as they do **not** provide short-circuit or overload protection. Instead, RCDs monitor the current passing through the device into the circuit (line) and the value coming back (neutral). If these values are different, some current must be escaping to earth. If the difference between the values on either side, i.e. line and neutral, is bigger than the residual operating current setting (I_N), the device will trip.

▼ **Figure 5.9** Typical RCD arrangement

RCDs should be rated in two ways:

(a) **nominal current:** this is based on the load current in the circuit as it is the amount of current that the device can safely switch, so in the case of a circuit with a load current of 17 A, an RCD with a nominal rating of 20 A would be used.

(b) **residual operating current setting ($I_{\triangle N}$):** this is the value of imbalance current, in milliamperes (mA), that causes the device to trip within 300 milliseconds (ms). RCDs are available in a range of residual operating current settings such as 10 mA, 100 mA, 300 mA, 500 mA and, most commonly, 30 mA.

▼ **Figure 5.10** Time/performance criteria for RCDs

TABLE 3A – Time/current performance criteria for RCDs to BS EN 61008-1 and BS EN 61009-1

RCD type	Rated residual operating current $I_{\triangle n}$ mA	Residual current mA	Trip time ms	Residual current mA	Trip time ms	Residual current mA	Trip time ms
General Non-delay	10	10		20		50	
	30	30		60		150	
	100	100	300 max.	200	150 max.	500	40 max.
	300	300		600		1500	
	500	500		1000		2500	

RCDs are found in many forms: they can be built into accessories such as socket-outlets and fused connection units or they can be incorporated into consumer units or mounted in separate enclosures. They could even be part of main switches for distribution boards or consumer units. This may be common where an installation forms part of a TT system but care must be taken to select the RCD rating accordingly as they must be capable for use as a main switch. They must also have a residual current rating that allows for discrimination because a single fault could cause the power in the entire installation that is protected by that main switch to be lost.

RCDs, unlike circuit-breakers, have a button on them labelled 'test'. If something mechanical is left in one position for any length of time, it may stick. As RCDs operate at very low current values, the energy generated may not be enough to trip a sticky mechanism.

Because of this, an RCD test button must be pushed regularly to keep the mechanism free from sticking. Circuit-breakers do not need this button as they operate at high current values, which will trip a sticky mechanism.

5.1.7 Residual current breaker with overload (RCBO)

Put simply, an RCBO is a circuit-breaker that also has an RCD built into it, giving it overload, short-circuit and residual earth current protection. RCBOs fitted for circuit protection require a neutral connection as well as the line because it needs to monitor the individual circuit current for imbalances.

RCBOs are able to meet the requirements of the IET Wiring Regulations in terms of additional protection and protection against overcurrent, as well as meeting the requirements for independent circuit control, meaning that a fault on one circuit does not affect another circuit by disconnecting it.

Like RCDs, an RCBO will have a test button that requires regular use, i.e. pressed every three months.

5.2 Selecting protective devices

▼ Table 5.2 Protective devices

Device Type	Advantages	Limitations	Common use
BS 88-3 fuse	A reasonable alternative to a rewireable fuse without having to change the consumer unit. High breaking capacity.	Fairly poor overload protection. Separate RCD needed to meet additional protection requirements. Requires replacement following rupture.	Used to be referred to as BS 1361 and are found in older domestic dwellings.
BS 88-2 gG & gM fuse	Very high breaking capacity. Able to select gM rating for motors or discharge lights.	Requires replacement following rupture. Separate RCD needed to meet additional protection requirements.	At the origin of installations that have a high prospective fault current. In industrial switch fuses.
BS 3036 fuse	It is inexpensive and has the ability to determine the nature of the fault as you can visibly see the broken fuse wire. Low breaking capacity.	Requires replacement following rupture. Separate RCD needed to meet additional protection requirements. Wrong rating of wire could easily be used.	Found in older domestic dwellings and some old commercial/industrial applications. Generally not used for new installations. BS 7671 still permits their use but certain conditions must apply, such as de-rating circuits and low prospective fault currents.

Device Type	Advantages	Limitations	Common use
BS EN 60898 circuit-breaker	Good overall circuit protection and easily reset following fault or overload. Good range of types depending on load.	Easily reset, meaning that ordinary persons could continually reset them, causing faults to worsen and shortening the life of the circuit-breaker. Have low short circuit capacity compared to other devices.	Found in all new installations, domestic, commercial and industrial.
BS EN 61009 RCBO	Ability to meet all the requirements of circuit protection.	Breaking capacities not suitable for very high prospective fault currents.	Where circuits require independent circuit protection meeting all requirements of BS 7671 (depending on prospective fault current values).
BS EN 61008 RCCB	Provides residual earth fault current protection where needed by special locations or additional protection. Can be installed remotely from an older consumer unit so an existing consumer unit can remain in place.	Does not provide short-circuit or overload protection.	Where additions or alterations are made to installations protected by fuses or as main or split-load protection.

5.3 Selectivity

In the IET Wiring Regulations, the term selectivity is used and defined as the ability of a protective device to operate in preference of another in series with it.

Good selectivity is important when designing an electrical installation that contains many protective devices. Selectivity essentially means that the protective device nearest to the fault will operate first. If a fault in an appliance's flexible cable happens, ideally, the fuse in the plug should blow before the circuit-breaker/RCBO that protects the circuit – this is good selectivity. If the circuit-breaker/RCBO operates first, power would be lost to all appliances on that circuit – an example of poor selectivity.

Equally, if a fault occurred in a circuit and the DNO's service fuse operated, poor selectivity would mean that the entire building would lose power.

Just because certain devices may have nominal current ratings (I_n) higher than another in the same circuit doesn't always mean that good discrimination has been achieved. The type of circuit-breaker also plays a part, as well as the type of fault.

Designers can check selectivity using the time/current graphs in Appendix 3 of the IET Wiring Regulations or provided by manufacturers.

Looking at the two examples below, the first graph shows that the characteristic lines for the local circuit-breaker and the distribution fuse do not cross at any point. This means that any fault in the circuit will trip the circuit-breaker first.

The second graph shows the local circuit-breaker line crossing the distribution fuse. This means that fault currents higher than the point where the lines cross will cause the distribution fuse to disconnect first.

▼ **Figure 5.11** Examples of selectivity

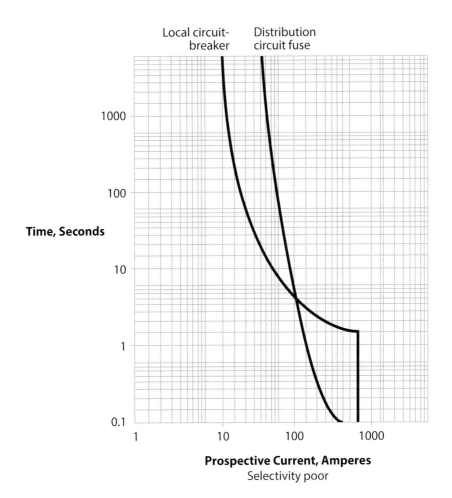

Selectivity poor

Selectivity may also need to be looked at where a circuit has more than one RCD or RCBO, regardless of their residual current settings, as the one closest to the supply may trip first.

To overcome this problem, the RCD nearest to the supply/furthest from the load should be a time-delayed or S-type RCD, meaning that it will only trip if the fault hasn't been cleared by the local RCD after a short duration in time. This time delay could be approximately 500 ms where the local RCD should disconnect within 300 ms.

5.4 Integration of devices and components

Regulation 536.4.203 requires that all devices and components installed in an assembly, for example, consumer unit or distribution board are compatible with one another. The reason for this requirement being introduced into the 18th Edition of BS 7671 is so that installers are aware that not all components are compatible with one another even if they from the same manufacturer and comply with the relevant product standards. If an installer decides to mix and match components (whether from the same manufacturer or a different one) within an assembly they should verify with manufacturers that all components are compatible with one another. The installer can carry out their own verification but this would ultimately mean that the installer will become the assembly manufacturer and any problems arising from the assembly as a result of incorrect integration of devices would be the responsibility of the installer/assembly manufacturer.

Imagine this situation: you need some new brake pads for your car. At the garage they don't have the specific ones required for your vehicle so they use some that are very similar, meet all the product standards requirements and are of good quality but are designed for a different make and model. You drive off and ten minutes down the road your brakes fail and you collide with another vehicle. When the insurance company investigate they find that the brake pads are not compatible with your vehicle. The responsibility will likely be with the mechanic who decided to use these particular pads instead of the ones that had been verified by the manufacturer.

5.5 Isolation and switching (Chapter 53)

A typical electrical installation will contain many different switches and/or isolators. It is very important to know what the purpose of the device is in order to select the right type. The IET Wiring Regulations require isolation and switching devices for the following reasons:

 (a) **isolation:** a means of cutting off electricity to all or part of an installation to allow safe electrical work to be carried out.

(b) **switching for mechanical maintenance:** a means of safely stopping a machine, piece of equipment or appliance from operating in any way to allow safe non-electrical work that does not involve exposure to potentially live parts.

(c) **emergency switching:** a means of quickly stopping a machine, piece of equipment or appliance from operating in the event of an emergency.

(d) **functional switching:** a means of controlling a machine, piece of equipment or appliance, such as turning it on or off.

(e) **undervoltage protection:** a means of stopping a machine, piece of equipment or appliance from automatically restarting following a loss or drop in voltage where sudden restarting could cause danger.

5.6 Isolators and switches (Chapter 46)

Before we look at the reasons for using a switch or isolator we need to understand the difference between them.

Isolators are rated for the current that can pass through them under normal load conditions but not switching ability. For example, a 100 A isolator could safely carry 100 A through it continually but it may not be capable of switching 100 A.

If an isolator is to be used, equipment should always be powered down first. This is because arcing (when electricity sparks as it travels through the air between terminals) takes place every time a device is opened under load. The amount of arcing depends on the type of load and some loads, such as inductive loads like motors, cause bigger arcs than resistive loads, such as immersion heaters or fluorescent lighting. This arcing could destroy device contacts or, as a minimum, create high resistance to the contacts, which in turn creates excessive heat through constant use.

If a device that is classed as an isolator is to be used, or positioned where it can be used, by ordinary persons as a switch (on-load), it must be de-rated in accordance with the manufacturer's instructions as a switch. For example, this means that a 100 A isolator may only be rated to switch 40 A.

Isolators are often categorised in the following ways:

(a) **SP-N (single pole & neutral):** opens the line conductor only but the neutral can be isolated by a link.
(b) **Double-pole:** opens the line and neutral conductors.
(c) **Triple-pole:** opens the three line conductors of a three-phase system and not the neutral.
(d) **TP-N (triple pole & neutral):** opens the three line conductors and the neutral can be isolated by a link.
(e) **Four-pole:** opens all three line conductors and the neutral.

5.7 Isolation

A means of isolating all live conductors must be provided for all installations, usually in the form of a main switch but other double pole devices may be used. The type of installation will affect the type of isolator used.

▼ **Table 5.3** Types of installation

Installation type	Supply	Earthing arrangement	Type of device as a minimum
Dwelling and light commercial	Single-phase	▶ TT ▶ TN-S ▶ TN-C-S	▶ Double-pole switch capable of full load switching. ▶ Must be rated to switch full load current.
Commercial/industrial	Three-phase	▶ TN-S ▶ TN-C-S	▶ TP-N device remote from CCU. ▶ Triple-pole within CCU.
Commercial/industrial	Three-phase	▶ TT	▶ Four-pole.

Isolation may also be achieved for individual circuits by the protective device for the circuit.

In all situations, the device used for isolation must be capable of being secured in the open or off position. For main-switches and circuit-breakers, this is normally achieved with a padlock and, if necessary, a special locking device. Where circuits are protected by fuses, the fuse must be capable of being fully removed and it should be kept by the person who undertakes the isolation; on many sites, there may be a 'permit to work' or 'lock-off' procedure to ensure that circuits or equipment are not unintentionally energised. Devices that block off the fuseway are available and prevent other fuses from being inserted whilst the circuit is securely isolated.

More guidance on safe isolation can be found in Section 2.

5.8 Switching off for mechanical maintenance

Mechanical maintenance covers many tasks and relates to a variety of different items of equipment. Providing that the work does not involve exposure to electrical terminals, switching off for mechanical maintenance can be done by ordinary persons provided they have received the necessary instruction from their employer. Mechanical maintenance includes, amongst many other tasks:

- **(a)** changing or adjusting drive belts on a machine;
- **(b)** installing or altering the pipework to a shower;
- **(c)** cleaning air filters inside an air handling unit (AHU); and
- **(d)** replacing lamps in a fluorescent luminaire.

For these situations, a device that is capable of switching full load current must be 'local' to the equipment. This means that, ideally, the device must be in a position it is in the control of the person carrying out the work and they can keep the switch under their supervision so that nobody else can switch it on.

Common devices for this purpose include:

- **(a)** fused connection units;
- **(b)** double-pole switches; and
- **(c)** plugs and socket-outlets.

If the device cannot be easily supervised then the device should be lockable, such as a lockable rotary switch, ensuring that the risk of someone else switching it back on has been removed; again, there may be a 'permit to work' or 'lock-off' procedure to ensure that circuits or equipment are not unintentionally energised.

5.9 Emergency switching

Emergency switching devices must be capable of cutting off the full load current.

In installations where items such as rotating machines can cause a danger, emergency switching is normally provided by stop buttons. When installing stop buttons to a machine or workshop, certain considerations need to be taken into account. For example, the stop button:

(a) must be near to the risk, so that someone using the machine can easily reach the button if they are in danger.
(b) must not be capable of being reset from another place unless the reset is by a key switch.
(c) should latch in the off position where (b) is not possible. This ensures that the person resetting the circuit has to be where the danger happened so that they can see that the danger has been removed before re-setting the circuit.
(d) should be key operated where the stop button could be reset by untrained persons, for example, a stop button in a school workshop.

5.10 Functional switching

This is probably the most common type of switch found in an installation. Functional switching allows equipment to be switched on or off as needed. It should be in a convenient place where the switch is operated manually. An example would be a light switch for a room, located by the entrance door to the room.

Where a room has more than one entrance, two-way and/or intermediate switching would normally be provided.

Functional switches may also be in the form of:

- **(a)** timeclocks;
- **(b)** passive infra-red (PIR) sensors;
- **(c)** plugs and socket-outlets (rated at or below 32 A);
- **(d)** switches on socket-outlets;
- **(e)** contactors;
- **(f)** standard light switches;
- **(g)** push switches; and
- **(h)** switched fused connection units.

Devices that should never be used as a functional switch include:

- **(a)** fuses;
- **(b)** luminaire connections, such as plug-in ceiling roses;
- **(c)** unswitched fused connection units; and
- **(d)** socket-outlets rated above 32 A.

Where a functional switch is controlling discharge luminaires, like fluorescent fittings, the switch must be rated high enough to manage the surge current that happens when the load is first switched on. Typically, a 6 A lighting circuit with fluorescent lights will have 10 A switches for this reason.

5.11 Undervoltage protection

Undervoltage protection is required where an unexpected restarting of a machine could cause danger.

Imagine a grinding wheel that is being used when a power cut occurs. If the grinding wheel does not have undervoltage protection, the wheel would dangerously and unexpectedly start spinning again once the power comes back on. This can be a huge risk to anyone near to or touching the grinding wheel.

A way to overcome this problem is to install a contactor with start and stop buttons next to the machine.

With this type of arrangement, the machine starts by pressing the start button and the contactor coil is energised, which in turn closes the contactor switch giving power to the machine. If the power fails, the contactor coil drops out, opening the circuit. If the power then comes back on, there is no danger as the contactor requires somebody to push the start button for the machine to come back on.

▼ **Figure 5.12** A typical motor circuit

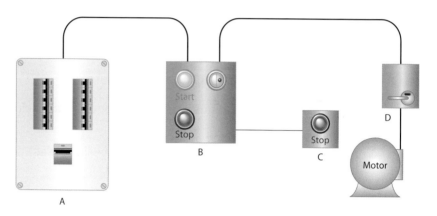

The diagram above shows a typical motor circuit, for example that is used to drive a mechanical pump. Each device labelled A to D performs the following function:

Device	Description	Function
A	Distribution board	■ Full isolation (main switch) ■ Circuit isolation (circuit breaker) ■ Protection (overcurrent)
B	Start/stop contactor with switch	■ Functional switching ■ Partial isolation ■ Emergency switching ■ Undervoltage protection ■ Overload protection
C	Remote stop button	■ Emergency switching
D	Local switch	■ Switching for mechanical maintenance ■ Isolation (as long as it is lockable)

	Test your knowledge
1.	What is an overload, as defined in Part 2 of the IET Wiring Regulations?
2.	Name a hazardous material that may be found in the carriers of BS 3036 fuses?
3.	What does the I_{cs} rating of a circuit-breaker relate to?
4.	Determine, using the graphs in Appendix 3 of the IET Wiring Regulations, the approximate time it takes for each of the following devices to disconnect with an overload of 60 A: (a) 32 A BS 88-3. (b) 32 A BS 88-2. (c) 30 A BS 3036. (d) 32 A BS EN 60898 type B.
5.	What is the maximum residual current rating (I_N) permitted for an RCD that provides additional protection in accordance with the IET Wiring Regulations?
6.	Which chapter from Part 5 of the IET Wiring Regulations covers the selectivity between RCDs?
7.	What specific requirements within Chapter 53 of the IET Wiring Regulations should be met where an 'off load' device for isolation is used within an installation?
8.	List three examples of common functional switches.
9.	What danger may exist if a machine didn't have suitable undervoltage protection?
10.	What would commonly be used to provide undervoltage protection for a machine?

5

Earthing and Bonding 6

After this section, you will have awareness of the following topics:

- ▶ What earthing is and why we need it
- ▶ Earthing arrangements
- ▶ What bonding is and why we need it
- ▶ Earth fault loop impedance

To understand how earthing works, you must first understand that electricity will take the easiest and quickest option every time. When it leaves one side of a transformer it wants to get to the other side as quickly and as easily as possible (the path of least resistance).

In the same way that if you want to drive a car from one side of a city to another and you had an option of whether to drive through the congested city centre or take the ring road (bypass) – it goes without saying that you would take the quickest route (the path of least resistance). Electricity is the same, well, sort of!

6.1 What is earthing and why do we need it?

The ESQCR states that the electricity distributor is, in most circumstances, obliged to provide the consumer with an earthing facility for a new supply to a low voltage installation. There may be safety reasons that prevent the distributor from providing an earth, for example, the neutral conductor may be susceptible to damage when used as a TN-C-S arrangement. An example of this is an overhead supply to a house in the countryside. If the neutral conductor is severed by a farmer in a tractor, 230 V can appear between true Earth and all metalwork within the installation. In this case, the supplier may refuse to provide a means of earthing and the consumer must then install a TT earthing arrangement instead. Ultimately, the consumer is responsible for ensuring that the installation is suitably earthed.

The purpose of earthing is to provide a path for current to flow through (in the event of a line to earth fault) so that the protective device operates within the required time (see Section 5 for more information on how protective devices work). For this to happen, the current needs to flow back to the neutral terminal of the distribution transformer (the origin), creating a complete circuit, albeit a complete fault circuit. To achieve this there needs to be a good conductive path from wherever the fault occurs back to neutral at the transformer via the main earth terminal at the consumer installation. The various ways in which this is achieved is called earthing.

6.1.1 Why do we call it earthing?

We call it earthing simply because in most electrical systems in the UK and many other countries, the neutral at the distribution transformer is connected to the general mass of Earth. By 'Earth' (note capital E!) we mean the thin layer of material that covers our planet, whether it be soil, mud, rock, clay or anything else you might find below your feet, whereas all the green and yellow conductors are connected to the earth (lower case 'e'!) of the electrical installation. That's right! There is a physical connection, by means of an earth rod and in most cases a copper mesh, between neutral and Earth in almost all distribution transformers in the UK.

When you remove the cover to the electrical terminals of most household appliances you will see that there is a cable identified with green/yellow insulation along with a cable with brown insulation and a cable with blue insulation . The green/yellow cable is the circuit-protective conductor (cpc) and connects the exposed-conductive-parts of appliances and the installation to earth. A circuit-protective conductor will carry fault currents to the consumer unit where the earthing conductor will send them on their merry way to Earth!

Chapter 54 of the IET Wiring Regulations provides the requirements for earthing arrangements and protective conductors.

> **Example: a light fitting**
>
> A light fitting has a metal body (exposed-conductive-part) that, in normal circumstances, is not live. But what if, when being moved around or accidentally knocked, the cable becomes damaged and the live conductor makes contact with the metal casing?
>
> Remember that electricity takes the easy route. In the case of this light fitting, imagine the filament of the lamp is a congested city: it is difficult for the electricity to pass through, and the metal casing is the motorway, a nice big chunk of conductive material to pass through.

When designing an electrical installation, the type of electrical supply from the DNO must be taken into consideration. This is especially important on a TT system where the general mass of the Earth is relied upon for fault current to flow around the earth fault loop impedance path. It is common for this type of system to rely on RCD protection due to the high earth fault loop impedance paths. The following figure shows how the cpc and earthing conductor provide a path for fault current to flow to the supply earthed neutral.

6.1.2 What is a fault to earth?

By this stage you will know that simple single-phase circuits consist of two live conductors: line and neutral. When one of these conductors, or when part of some equipment that carries current, becomes damaged and connects with an exposed-conductive-part (see Section 1) it creates a path for the current to flow to the cpc. This cpc is connected between the exposed-conductive-parts and the main earthing terminal (MET) of that installation. If this offers the path of least resistance then the bulk current flow will divert and travel through the cpc back to the earthed neutral of the distribution transformer (source). As long as the installation meets the requirements of the IET Wiring Regulations, this should be sufficient current to flow, which will operate the protective device within the required time.

6.1.3 What happens if there is bad connection to earth or no connection at all?

Keep in mind that installations in the UK have the neutral conductor connected to the general mass of the Earth at the distribution transformer and that the current will travel through the general mass of the Earth to get back to the transformer if it is the easiest path.

Now imagine an installation that is not correctly earthed, such as a corroded protective conductor (creating higher resistance) or an installation that is not earthed at all (no path). If the same fault was to occur and the exposed metallic part of the equipment became live, it would stay live until something or someone created a path for the current to flow to earth or through the general mass of the Earth. This may happen if a person was to touch a metal part that has become live, as a result of a fault, while standing on the ground (Earth) or touching a piece of conductive material that emerges from the ground, such as metallic pipework or structural steel (for more information about extraneous-conductive-parts see Section 1). The consequences have proven fatal on many occasions in the past.

6.1.4 Earthing arrangements

Terre neutral-connected-separated (TN-C-S) arrangement

For more information about terre neutral-connected-separated (TN-C-S) arrangements, see Regulation 542.1.2.2 in the IET Wiring Regulations.

At the distribution transformer, the neutral conductor is connected to Earth by means of an earth electrode. The neutral is now 'earthed' and the term Terre, which is French for earth, describes that the source is earthed. From the transformer, a single conductor performs both the neutral and earthing functions, known as the protective earthed neutral (PEN conductor). Along with the line conductor, the PEN conductor forms part of a composite cable which is buried in the ground and is run to the consumer's property. At this point, the arrangement is TN-C, the C describes that the earthing and neutral functions are combined in the single conductor. When the cable enters the property a special distribution fuse carrier has an extra terminal, often referred to as the PME terminal, which provides an earthing connection for the electrical installation.

You can see from Figure 6.1 that the TN-C-S arrangement uses the neutral supply conductor (PEN) as the means of earthing for the installation. This means that if a line to earth fault were to occur within this installation, for example, a faulty kettle, the fault current will travel through the circuit-protective conductor to the main earthing terminal of the installation. From there, the installation's earthing conductor, usually a 16 mm² green and yellow insulated copper conductor, will carry the fault current to the PME earthing terminal. From there, the fault current will travel through the combined PEN conductor (the blue one) and then back to the transformer. This flow of current will be sufficient (if the installation complies with the IET Wiring Regulations) to operate the protective device in the required time.

▼ **Figure 6.1** TN-C-S arrangement

Terre Neutral-Separated (TN-S) arrangement

For more information about terre neutral-separated (TN-S) arrangements, see Regulation 542.1.2.1.

Terre is connected to the neutral conductor at the distribution transformer as it is with the TN-C-S arrangement. However, on the TN-S arrangement the earth conductor from the transformer to the property is separate. The separate conductor is usually the steel wire armour on the supply cable that in most cases provides adequately low resistance.

You can see from Figure 6.2 that the TN-S arrangement uses a separate conductor as a means of earthing the installation. This means that if a line to earth fault were to occur on this installation, for example, a faulty kettle, the fault current will travel through the cpc to the main earthing terminal of the installation down to where the supply cable enters the property. After this it travels through a separate protective earth conductor back to the transformer. This flow of current will be sufficient (if the installation complies with the IET Wiring Regulations) to operate the protective device in the required time.

▼ **Figure 6.2** TN-S arrangement

Terre terre (TT) or earth earth arrangement

For more information about terre terre (TT) or earth earth arrangements see Regulation 542.1.2.3.

At the distribution transformer, the neutral conductor is connected to Earth by means of an earth electrode. The neutral is now 'earthed' and the term Terre, which is French for (Earth) describes that the source is earthed. This connection of the neutral conductor to earth at the distribution transformer as it is with the TN-C-S and the TN-S arrangements. However, with a TT earthing arrangement, the earth conductor from the transformer to the property is actually the general mass of the Earth! As a result, to create a path for current to flow from a consumer's installation to the transformer, an earth electrode is driven into the ground close to the consumer's installation. This provides a good return fault path via which the current can flow to the general mass of the Earth. From here it will flow through the ground to the distribution transformer.

However, the general mass of Earth is not the most effective conductor and varies significantly depending on location and moisture content in the ground.

You can see from Figure 6.3 that the TT arrangement uses the general mass of Earth as a means of earthing the installation. This means that if a line to earth fault were to occur on this installation (for example, our faulty kettle) the fault current will travel through the circuit protective conductor to the main earthing terminal of the installation, along the earthing conductor, down to the earth rod then through the general mass of the Earth back to the distribution transformer. This flow of current will be sufficient (if the installation complies with the IET Wiring Regulations) to operate an RCD in the required time.

▼ **Figure 6.3** TT arrangement

6.2 Protective equipotential bonding

Section 544 of the IET Wiring Regulations covers the requirements for protective equipotential bonding conductors, including information relating to:

  (a) the required csa of the main protective bonding conductors and supplementary bonding conductors;
  (b) where and how they should be connected in relation to the consumer unit; and
  (c) where metallic pipework enters the building, etc.

Section 4 of the *On-Site Guide* provides comprehensive guidance about the installation of bonding conductors.

6.2.1 What is equipotential bonding and why do we need it?

In an installation where the protective measure is automatic disconnection of supply (ADS), where the protective device, i.e. the RCD, detects a fault and disconnects the supply to the circuit automatically, protective equipotential bonding must be provided in accordance with the IET Wiring Regulations. The purpose of main protective equipotential bonding is to reduce touch voltages during an earth fault that may exist between:

(a) extraneous-conductive-parts (metallic parts that do not form part of the installation but may become live in the event of an earth fault, such as structural steel and metallic pipework) and other extraneous-conductive-parts.

(b) extraneous-conductive-parts and exposed-conductive-parts.

(c) exposed-conductive-parts and Earth.

6.2.2 Potential difference

One thing to remember is that the general mass of the Earth is considered to have zero potential or zero volts. This means that the difference between, let's say, a single-phase line conductor and the general mass of the Earth will be 230 V so the potential difference is 230 V. If something was to create a path between the two it would have the full 230 V placed across it, whether it be a piece of conductive material or a human body.

> **Something you should never try**
>
> I remember that, when I was a child, I was always wondering why birds could fly on and off bare electrical transmission cables without being electrocuted. People would generally explain to me that it was because they were not touching two wires at the same time. While this is true to some extent, it did not become clear to me until I had a better understanding of how electricity actually works.
>
> Whether it's a bird or even me hanging onto an overhead cable, the principles are the same. If I were to hang on to one part of the cable with my right hand and another part of the same cable with my left hand I would be able to hang there quite happily, well, apart from being 20 or so metres above the ground. This is because the potential difference between the two parts of the cable in my hands would be zero.
>
> This can be proven with a voltmeter by holding the two probes onto the same live conductor a short distance apart (it is recommended that this is done under the supervision of a lecturer and on an extra low voltage system). The value shown will be zero or at least very close to zero. This is, essentially, what we are trying to achieve when we install equipotential bonding: to achieve as close to zero potential as possible between conductive parts.

Hopefully this helps when trying to understand 'equipotential'. It essentially means 'equal potential', i.e., to have the same potential at two or more points. Although this is almost impossible to achieve, any reduction in the potential difference between two parts is an improvement to the safety of the installation during an earth fault.

6.2.3 What happens during a fault?

During a fault to earth, the current will flow around the earth fault loop impedance path until it reaches a sufficient magnitude which causes automatic operation of the protective device. However, if something or someone were to come into contact with a live exposed-conductive-part that wasn't earthed and an extraneous-conductive-part at the same time and the installation had insufficient protective equipotential bonding, they could immediately create a path for current to flow through the body between these two parts. The voltage between these two parts is known as 'touch voltage' (U_t).

Touch voltage can give an electric shock, which could cause serious injury or even take someone's life. Equipotential bonding, in accordance with the IET Wiring Regulations, can reduce touch voltage during an earth fault to an acceptable level (low enough to prevent a serious risk from electric shock) until the protective device operates.

6.2.4 How does protective equipotential bonding work?

Remember that electricity will always take the easiest path back to the neutral at the supply transformer. If the easiest path is through the cpc and the protective equipotential bonding conductors, then that's where most of the current will flow. If there is an alternative route, such as through a person to an extraneous-conductive-part, then that person is going to receive a nasty shock. This is why it is important not only to provide protective equipotential bonding but to ensure that the csa of the conductor is sufficient enough to provide an easy path through which current can flow.

In the figures below, a line to earth fault has occurred on a TN-C-S system and has introduced a voltage potential between the exposed-conductive-part of the electrical equipment and the extraneous-conductive-part (metal pipework). The fault path (earth fault loop) is shown by the purple dotted line.

In Figure 6.4 there is no protective equipotential bonding. When an earth fault occurs, the touch voltage may be quite high between the exposed-conductive-part (the metallic casing of the lamp) and the extraneous-conductive-part (the metallic pipework rising from the ground). This is because the man in the diagram is creating an easy path for the electricity to flow back to the supply transformer via the metal pipework and the general mass of the Earth. In this example, there is a touch voltage of 143 V, which is travelling through the person. This has the potential to cause serious injury or even death.

▼ **Figure 6.4** Installation without protective equipotential bonding

In Figure 6.5 the installation has protective equipotential bonding, which connects the extraneous-conductive-part as it enters the property to the main earthing terminal at the consumer unit. By installing this additional conductor, the potential difference that exists during a fault to earth is reduced. This will significantly lessen the magnitude of the touch voltage that can exist between the exposed-conductive-part and the extraneous-conductive-part of the installation.

In Figure 6.5, we can see that the touch voltage has been reduced to 50 V.

▼ **Figure 6.5** Installation with protective equipotential bonding

6.2.5 Connecting two or more extraneous-conductive-parts using a single conductor

If a single protective bonding conductor is used to bond two or more extraneous-conductive-parts, it should be one continuous length of cable so that, in the event that any part of the bonding group becomes detached from the extraneous-conductive-part, there will still be continuity between any other extraneous-conductive-parts and the MET.

Guidance on the correct size of bonding conductors to be used can be found in Table 4.4(i) and 4.4(ii) of the *On-Site Guide*. The minimum size that should be used in any installation is 6 mm. The size of the main protective bonding conductor is directly related to the size of the incoming live conductors.

In Figure 6.6 an example of how the conductor should be connected to the terminal (unbroken) is shown.

▼ **Figure 6.6** Unbroken bonding conductor

6.3 Supplementary equipotential bonding

The principles of supplementary equipotential bonding are the same as equipotential bonding. The only difference is that it further reduces any potential voltage in a particular part of an installation during an earth fault that may arise between exposed-conductive-parts and extraneous-conductive-parts.

There are certain circumstances where it is required by the IET Wiring Regulations to have supplementary equipotential bonding, particularly in installations and locations where there is an increased chance of electric shock, such as locations containing swimming pools. Part 7 of the IET Wiring Regulations provides more information on the zones that require this additional protection.

Section 4 of the *On-Site Guide* provides information about how to install protective equipotential bonding. Table 4.6 is particularly useful in identifying the required csa of the bonding conductors that should be used to achieve the level of protection required.

All the requirements and guidance available in the IET Wiring Regulations and associated guidance for protective equipotential bonding are intended to help you to reduce the voltage between parts that may exist during an earth fault and, ultimately, reduce the severity of electric shock that someone may experience.

6

	Test your knowledge
1.	Is the earth conductor of an electrical installation a live conductor?
2.	If the supplier refuses to provide a connection to earth at the consumer's installation, what type of earthing arrangement would be considered?
3.	The purpose of earthing is to provide a path for current to flow to which terminal at the distribution transformer in the event of a fault so that the protective device will operate?
4.	What is used to connect the neutral side of the transformer to the general mass of the Earth?
5.	Write the full description of the following abbreviations: (a) TN-S. (b) TN-C-S. (c) TT.
6.	Is the metal casing of a light fitting an extraneous-conductive-part or an exposed-conductive-part?
7.	Is a structural reinforced steel girder rising from the ground an exposed-conductive-part or an extraneous-conductive-part?
8.	When would a touch voltage be present?
9.	When the two probes of a voltmeter are connected to the same energised line conductor what would be the expected reading?
10.	Why is it required to have one continuous conductor for protective equipotential bonding when it is connected to more than one extraneous conductive part?

Cable Calculations 7

After this section, you will have awareness of the following topics:

- ▶ Why we need cable calculations
- ▶ Current rating of equipment, protective devices and current-carrying capacity of cable
- ▶ Rating factors
- ▶ Volt drop

7.1 Why do we need cable calculations?

When a circuit is in use there is current travelling through the conductors. If lots of appliances are being used, for example, a washing machine, a hair dryer etc., or more power is needed to operate the appliance (such as a supply for a heater or an oven) there will be more current flowing through the cable.

Cables have a limit! If the installation hasn't been designed correctly and the cable for the circuit is too small (i.e. it does not have the required current-carrying capacity) there is a high risk of the cable overheating and causing a fire. This could ultimately lead to the destruction of property and present a serious risk to life.

It is therefore necessary to consider what the circuit is going to be used for, the environment where the cable will be installed and the type of cable used. As long as the installation meets the requirements of the IET Wiring Regulations, the cable should be adequate to carry the design current under the specified installation conditions.

7.2 Current rating of equipment, protective devices and current-carrying capacity of cable

7.2.1 Current ratings and capacities

There are various symbols and abbreviations that you need to be familiar with to efficiently calculate the required size of a cable for an installation. Below, we discuss the different types of current ratings. You'll notice that each begins with the 'I', which is the SI symbol for current.

I_b (design current)

This is the current that will be used by an item of electrical equipment under normal operating conditions. For example, if you have a **1,000 W** heater supplied by **230 V**, I_b will be the maximum current that will flow to operate the appliance. In this example the design current will be **4.35 A** (see how to work this out later in this section). For circuits designed to supply multiple appliances, such as cooker circuits, diversity should be taken into consideration (guidance on this is given in Section 12).

I_n (rating of protective device)

This is the rating of the device that is designed to operate and break a circuit should a fault or overload occur.

I_t (tabulated current-carrying capacity of cable)

This is the current-carrying capacity of the cable without the application of any correction factors or installation conditions taken into consideration essentially, the maximum current a cable will be able to carry (see Appendix 4 tables 4D1A through 4D5 of the IET Wiring Regulations).

I_z (current-carrying capacity of a cable for continuous service)

This is the current-carrying capacity of the cable for continuous service with all necessary correction factors applied and installation conditions taken into account, for example, if the cable has a long distance to travel and it has to pass through thermal insulation, factors need to be applied and its current-carrying capacity will be lower than that of the specified value in Appendix 4 tables 4D1A through 4D5 of the IET Wiring Regulations.

7.2.2 Rating factors to consider in calculations

What are correction factors?

The resistance of a cable is affected by heat. As the cable gets warmer the resistance increases. The retention of heat in a cable is affected by many factors. Listed below are some of the factors that should be considered when designing an installation.

C_a (ambient temperature)

This is the temperature of the environment in which the cable will be installed.

▼ **Figure 7.1** Temperature of environment

C_c (cables buried in the ground)

When cables are buried in the ground, the heat generated is not able to readily dissipate and the temperature of the conductors will rise causing the resistance of the conductors to increase when the circuit is in use.

▼ **Figure 7.2** Cables buried in the ground

C_d (depth of burial)

Heat dissipation will vary depending on the depth at which the cable is buried.

▼ **Figure 7.3** Depth of the cables buried in the ground

C_f (semi-enclosed fuses – see BS 3036)

Where a semi-enclosed fuse to BS 3036 is used to protect a cable against overload, the cable needs to be derated by 0.725 in addition to any other necessary factor; this is one of the disadvantages of using this type of protective device.

▼ **Figure 7.4** BS 3036 Fuses

C_g (grouping)

Various influences affect cables when they are grouped together. The build-up of heat has the most impact on the resistance of conductors.

▼ **Figure 7.5** Cable grouping

C_i (thermal insulation)

Thermal insulation, common in most installations, can cause cables to overheat when installed in or alongside it.

Figure 7.6 Cable run through thermal insulation

C_s (soil resistivity)

The resistance of soil can vary depending on the location, moisture content etc.

Figure 7.7 Cable run underground

> **Remember ...**
>
> To help you remember these, try to associate the letters with the correction factors, for example:
>
> ▶ C_a = Rating factor **A**mbient temperature (see Table F1 of the IET's *On-Site Guide*
> ▶ C_c = Rating factor Buried, think of it as **C**overed
> ▶ C_d = Rating factor **D**epth of burial
> ▶ C_f = Rating factor **F**use (BS 3036 only)
> ▶ C_g = Rating factor **G**rouping (see Table F3 of the IET's *On-Site Guide*)
> ▶ C_i = Thermal **i**nsulation (see Table F2 of the IET's *On-Site Guide*. Note for cables installed using methods described in Tables F4(i), F5(i) and F6 the rating factor that needs to be used is 1 as the allowance for thermal effects has already been taken into account).
> ▶ C_s = **S**oil resistivity

7.3 How design current, protective device rating and current-carrying capacity relate to each other

7.3.1 Design current, protective device rating and cable rating

I_b **(design current)** should always be equal to or less than I_n.

I_n **(over-current rating of protective device)** should always be equal to or less than I_t.

I_t **(tabulated current-carrying capacity of a cable)** will invariably be lower than I_z.

I_z **(current-carrying capacity of cable for continuous use under particular installation conditions)**, providing the value is equal to or greater than the rating of the protective device (I_n), the cable will be suitably protected.

This can be viewed as:

▶ $I_b \leq I_n \leq I_t$
▶ $I_n \leq I_z$

If I_b was higher than the rating of the protective device it would cause the device to operate whenever the circuit was in use and would, therefore, be impractical.

If the protective device was rated higher than I_z there would be a risk of the cable overheating and catching fire before the protective device could operate.

7.3.2 Identifying the values to be used when calculating the required rating of cable and protective device

Identifying the supply voltage

In most UK domestic dwellings, the supply voltage will be 230 V single-phase. Note that the supply voltage at the terminals of an item of equipment is very rarely 230 V when we measure it. Many factors affect the supply voltage, such as the length of run of the cables or electrical loading in the area. The ESQCR permits the supply voltage to fluctuate between -6 % and +10 %, which means that when we measure the voltage at the supply terminals on an electrical installation we would expect the voltage to be between 216.2 V or 253 V. With this in mind, we base all calculations on a nominal voltage of 230 V. The voltage and number of lines that are needed to operate a particular piece of equipment must be determined at the design stage of the installation.

Identifying the power rating of the appliance or equipment

Most appliances or equipment will have voltage and power ratings clearly identified on them and should also be stated in the manufacturer's instructions.

Once both the voltage and the power ratings have been identified you can then work out the design current to determine the necessary size of cable, using the following formula.

Working out the design current

For example, if the installation consisted of one storage heater with a power rating of **3 kW**, that is what will be used in the calculation. If there are three storage heaters at **3 kW** each, the total that will be used in the calculation will be **3 × 3 kW (9 kW)**.

$$I_b(\text{Design I}) = \frac{P(\text{PowerW})}{V(\text{Volts})}$$

In this example the equation would look like this:

$$I_b(\text{Design I}) = \frac{9000\,(9\text{ kW})}{230} = 39.1\text{A}$$

Determine the appropriate protective device rating

Once the design current has been calculated, the appropriate protective device rating (I_n) can be selected, which is always equal to or greater than I_b.

Determine the size of the cable

By determining I_t using the tables provided in Appendix 4 of the Wiring Regulations a cable with sufficient current-carrying capacity can be found. There are many different installation methods that help determine the required cross-sectional area of the conductor. A description of each method can be found at the top of each table.

Once I_n has been selected the requirements for I_t can be identified, using the formula:

$$I_t \geq \frac{I_n}{C_a \times C_c \times C_d \times C_f \times C_g \times C_i \times C_s}$$

This is assuming that there are seven factors to take into consideration. If, for example, you have no thermal insulation to run cables through, you would simply remove C_i from the calculation.

If the only factor to take into consideration was the ambient temperature C_a, it would be the only factor in the calculation, as shown below:

$$I_t \geq \frac{I_n}{C_a}$$

7

> **Example**
>
> A circuit needs to be installed to supply a 230 V appliance with a power rating of 5 kW. The ambient temperature in the room where it is to be installed is 35 °C. The 70 °C thermoplastic flat sheathed cable will need to be run through 2.5 m of thermal insulation exceeding 100 mm in thickness in a stud wall, touching the inner wall. The total length of the cable run is 10 m.
>
> The calculation is as follows:
>
> To determine the design current (I_b):
>
> $$I_b = \frac{P}{V}$$
>
> P = 5 kW (5,000 W)
>
> V = 230 V
>
> $$I_b = \frac{5,000 \, W}{230 \, (V)} = 21.73 \, A$$
>
> Now that the design current has been calculated, the other factors can be used to work out the size of the cable required. We already have one value:
>
> I_b = 21.73 A
>
> As previously stated, the protective device (I_n) must be rated either equal to or higher than the design current (I_b), so it should be the lowest rating available:
>
> 21.73 A ≤ I_n
>
> This would mean using a protective device with a rating of 21.73 A or higher. For this example a circuit-breaker (BS EN 60898) with a rating of **32 A** is used.
>
> The first value we have is: I_n = **32**

The next value that can be used is the correction factor for the ambient temperature (C_a). This can be obtained using Table 4B1 from Appendix 4 of the IET Wiring Regulations and Table F1 from the *On-Site Guide*. The cable being used is flat 70 °C (the maximum operating temperature of the cable) thermoplastic and the ambient temperature is 45 °C, so the factor to be used is 0.94.

▼ **Figure 7.8** Checking Table 4B1 from Appendix 4 of the IET Wiring Regulations

TABLE 4B1 – Rating factors (C_a) for ambient air temperatures other than 30 °C

Ambient temperature[a] °C	Insulation				
	60 °C thermosetting	70 °C thermoplastic	90 °C thermosetting	Mineral[a]	
				Thermoplastic covered or bare and exposed to touch 70 °C	Bare and not exposed to touch 105 °C
25	1.04	1.03	1.02	1.07	1.04
30	1.00	1.00	1.00	1.00	1.00
35	0.91	0.94	0.96	$C_a = 0.94$	0.96
40	0.82	0.87	0.91	0.85	0.92
45	0.71	0.79	0.87	0.78	0.88
50	0.58	0.71	0.82	0.67	0.84
55	0.41	0.61	0.76	0.57	0.80
60		0.50	0.71	0.45	0.75
65		–	0.65	–	0.70
70		–	0.58	–	0.65
75		–	0.50	–	0.60
80		–	0.41	–	0.54
85		–	–	–	0.47
90		–	–	–	0.40
95		–	–	–	0.32

a For higher ambient temperatures, consult manufacturer.

7

As we can see from Table 4B1, $C_a = 0.94$.

The next value to be calculated is the thermal insulation factor C_i. This can be obtained using Table 4D5 from the IET Wiring Regulations or F6 from the IET *On-Site Guide*, where the current-carrying capacity of the cable can be found. This table shows the various different installation methods and how they affect the current-carrying capacity of the cable. As described earlier in this section, if the installation method can be found in this table the rating factor to be used in the calculation is 1.

Now that all possible values and factors have been obtained the calculation for I_t is:

$$I_t \geq \frac{I_n}{C_a}$$

$$I_t \geq \frac{32}{0.94}$$

$$I_t = 34.04$$

Using the value 34.04 A, refer to Table 4D5 to identify the size of the cable to be used. The installation method, as mentioned earlier, is 102.

TABLE 4D5 – 70 °C thermoplastic insulated and sheathed flat cable with protective conductor (COPPER CONDUCTORS)

CURRENT-CARRYING CAPACITY (amperes) and VOLTAGE DROP (per ampere per metre):
Ambient temperature: 30 °C
Conductor operating temperature: 70 °C

Conductor cross-sectional area	Method 100# (above a plasterboard ceiling covered by thermal insulation not exceeding 100 mm in thickness)	Method 101# (above a plasterboard ceiling covered by thermal insulation exceeding 100 mm in thickness)	Method 102# (in a stud wall with thermal insulation with cable touching the inner wall surface)	Method 103# (in a stud wall with thermal insulation with cable not touching the inner wall surface)	Reference Method C* (clipped direct)	Reference Method A* (enclosed in conduit in an insulated wall)	Voltage drop (per ampere per metre)
1	2	3	4	5	6	7	8
(mm²)	(A)	(A)	(A)	(A)	(A)	(A)	(mV/A/m)
1	13	10.5	13	8	16	11.5	44
1.5	16	13	16	10	20	14.5	29
2.5	21	17	21	13.5	27	20	18
4	27	22	27	17.5	37	26	11
6	34	27	35	23.5	47	32	7.3
10	45	36	47	32	64	44	4.4
16	57	46	63	42.5	85	57	2.8

So you can see that, based on the information provided in this section and using the table above, a 6 mm² can be used for this installation providing the voltage drop is within the required percentage.

7.4 Voltage drop

7.4.1 What is voltage drop?

Regulation 525 of the Wiring Regulations outlines the requirements for Voltage drop and Table 4Ab from Appendix 4 provides the maximum permissible voltage drop for consumer installations.

When current travels through cable, energy is being used.

Here is a scenario that can be related to the energy used when electricity is travelling through a conductor (the 'tunnel analogy').

If a person lifts an object next to them weighing 1/3 of their body weight they would, in most cases, not find this difficult.

If they were to crawl through a 20 m long tunnel only just big enough to fit through (high resistance), and then lift an object 1/3 of their weight, it would be considerably more difficult. This is because they will have already used a lot of energy to get to the end of the tunnel, leaving little energy for lifting the weight.

If they walked through a large tunnel (low resistance) they would have more energy left to lift the weight at the other end.

A similar principle can be associated with cable. Every millimetre that is travelled uses energy. The longer the cable run, the higher the resistance and the more energy needed to get to the end.

In other words, a larger cable has less resistance and is easier to travel through. Also, in the event of a fault, sufficient current will flow to ensure that the protective device will operate within the required time.

Where there is a nominal voltage of 230 V public supply:

    (a) for lighting, the voltage drop is to be no more than 3 % (6.9 V).
    (b) for other circuits, the voltage drop is to be no more than 5 % (11.5 V).

Where there is a nominal voltage of 230 V private supply (i.e. from a privately owned transformer):

    (a) for lighting, the voltage drop is to be no more than 6 % (13.8 V).
    (b) for other circuits, the voltage drop is to be no more than 8 % (18.4 V).

Each cable has a millivolts per amp per metre (mV/A/m) factor. This tells us the voltage drop per amp for every metre in length, so we can calculate the overall voltage drop on a particular circuit. To establish whether the 6 mm² cable calculated earlier is sufficient, we need to know the (mV/A/m) for the cable in question. This value can be found in Table 4D1B through to Table 4J4B of the IET Wiring Regulations.

For thermal insulated and sheathed flat cable with protective conductor, Table 4D5 is used.

TABLE 4D5 – 70 °C thermoplastic insulated and sheathed flat cable with protective conductor (COPPER CONDUCTORS)

CURRENT-CARRYING CAPACITY (amperes) and VOLTAGE DROP (per ampere per metre):

Ambient temperature: 30 °C
Conductor operating temperature: 70 °C

Conductor cross-sectional area	Method 100# (above a plasterboard ceiling covered by thermal insulation not exceeding 100 mm in thickness)	Method 101# (above a plasterboard ceiling covered by thermal insulation exceeding 100 mm in thickness)	Method 102# (in a stud wall with thermal insulation with cable touching the inner wall surface)	Method 103# (in a stud wall with thermal insulation with cable not touching the inner wall surface)	Reference Method C* (clipped direct)	Reference Method A* (enclosed in conduit in an insulated wall)	Voltage drop (per ampere per metre)
1	2	3	4	5	6	7	8
(mm²)	(A)	(A)	(A)	(A)	(A)	(A)	(mV/A/m)
1	13	10.5	13	8	16	11.5	44
1.5	16	13	16	10	20	14.5	29
2.5	21	17	21	13.5	27	20	18
4	27	22	27	17.5	37	26	11
6	34	27	35	23.5	47	32	7.3
10	45	36	47	32	64	44	4.4
16	57	46	63	42.5	85	57	2.8

Once **mV/A/m** has been determined, the design current (I_b) is added to the calculations.

Finally, the distance in metres (**L**) must be identified.

The equation looks like this:

$$\text{volt drop}(V) = \frac{mV/A/m \times I_b \times L}{1000}$$

If the total length of cable is 10 m the equation will look like this:

$$\text{Total volt drop} = \frac{11 \times 10.869 \times 10}{1000} = 1.19$$

As this is within the 5 % allowance for voltage drop, a 6 mm² flat, twin and earth sheathed cable can be used for this installation.

Remember...

To recap, here are the steps for following the correct cable calculation procedure.

Step 1: Identify the voltage and power requirements.

Step 2: Use the calculation $I_b = \frac{V}{P}$ to identify the design current.

Step 3: Now you can identify the required rating of the protective device (I_n) using tables from Section 7 of the *On-site Guide*. Remember that it must be equal to or higher than the design current.

7

> **Remember...**
>
> **Step 4:** Determine the size of the cable to be used. You can find current-carrying capacities of various cables in Appendix 4 of the IET Wiring Regulations.
>
> **Step 5:** Establish the voltage drop.

To calculate the voltage drop of a particular cable you will need to look through Appendix 4 of the IET Wiring Regulations. Below the tables that show you current-carrying capacity there is a table that contains the **mV/A/m** for each cable in particular installation conditions. For example, if Table 4D1A shows the current-carrying capacity for single core 70 °C thermoplastic insulated cables non-armoured, with or without sheathing, then Table 4D1B will show the **mV/A/m** for the same type of cable and installation condition.

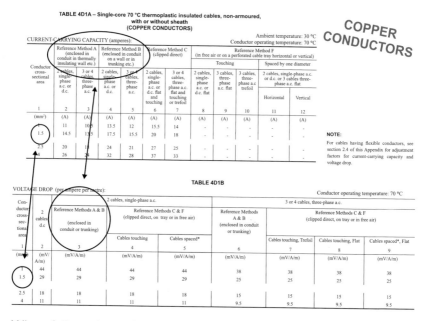

When followed step by step the calculations are simple. The trick is to not complicate it. Complete one step before moving on to the next.

Test your knowledge

The following examples are intended as examples to practice calculating the required size of the live conductors only.

1. A customer would like to have an electric shower installed in their upstairs bathroom. Take a look at the following information and work out the appropriate size of 70 °C thermoplastic sheathed flat twin & earth cable and protective device required. The cable will have to run alongside the thermal insulation touching the inner wall. The protective device to be used should comply with BS EN 60898 the ambient temperature of the room is expected to be 35 °C the measurements given are the total length of the cable run.

   (Use tables in Appendix F of the IET's *On-Site Guide* or tables from Appendix 4 of the IET Wiring Regulations; pay particular attention to the type of cable that is to be used.)

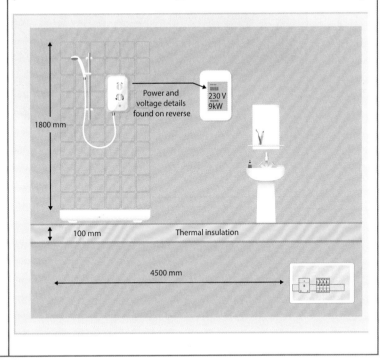

Test your knowledge

2. A customer requests a power supply to their shed for a light. They would like to have a 500 W security light on the outside.

   Using the information below calculate the size of the cable required to supply the halogen security light.

   The cable to be used should be three core armoured 70 °C thermoplastic insulated buried direct in the ground, the ambient temperature is not expected to exceed 20 °C and the depth of burial is 0.5 m with a soil resistivity of 2.5 K.m/W.

Final circuits 8

After this section, you will have awareness of the following topics:

- Designing and installing a new circuit
- Consumer unit arrangements
- Luminaires and lighting installations radial final circuits
- Ring final circuits
- Spurs
- Shower circuit
- SELV
- PELV
- FELV

8.1 General installation practices

There are certain requirements that must be complied with when making any additions or alterations to an existing electrical installation. However, just because an existing installation doesn't meet all the current requirements as set out by the IET Wiring Regulations does not always mean that the installation needs upgrading. For example, if an older installation has red and black insulation around the conductors instead of brown and blue, it does not mean that you cannot carry out any work on the installation without replacing all the cable in the system first. This would be impractical and most likely upset the customer having to pay the bill.

8

Just because it's old doesn't mean it's unsafe. You need only upgrade what is necessary to maintain the safety requirements of an electrical installation.

As an electrician, you will need to be satisfied that all the fundamental safety aspects of the installation are sufficient, for example, that protective devices are in good working order and that the disconnection times can be achieved. Using the guidance available, it is the responsibility of the skilled person carrying out the inspection or installation work on an electrical system to determine if the installation can be deemed fit for continued service.

The following text can be found in the introduction of the IET Wiring Regulations:

The Regulations apply to the design, erection and verification of electrical installations, also additions and alterations to existing installations. Existing installations that have been installed in accordance with earlier editions of the Regulations may not comply with this edition in every respect. This does not necessarily mean that they are unsafe for continued use or require upgrading.

8.2 Designing and installing a new circuit

Designing a circuit can be a long, difficult task that involves a great number of considerations that usually consist of many calculations. Thankfully, the IET provides valuable information in the *On-Site Guide* and the Guidance Note suite that have already worked out many of the calculations for you, which will not only save you a lot of time but will also provide you with the necessary guidance so that the electrical systems you are installing will meet the requirements of the IET Wiring Regulations.

Example

The maximum earth fault loop impedance values need to be obtained to ensure that a protective device operates within the required disconnection times. Below is an excerpt from Appendix B of the *On-Site Guide* showing the predetermined earth fault loop impedance values:

Circuit-breaker type	Circuit-breaker rating (amperes)												
	3	5	6	10	15	16	20	25	30	32	40	45	50	63	100
1	14.56	8.74	7.28	4.4	2.93	2.76	2.2	1.76	1.47	1.38	1.1	0.98	0.88	0.7	0.44
2	8.4	5.0	4.2	2.5	1.67	1.58	1.25	1.0	0.83	0.79	0.63	0.56	0.5	0.4	0.25
B	11.65	7.0	5.87	3.5	2.3	2.2	1.75	1.4	1.17	1.1	0.88	0.78	0.7	0.56	0.35
3&C	5.82	3.49	2.91	1.75	1.16	1.09	0.87	0.7	0.58	0.55	0.44	0.38	0.35	0.27	0.17

For a more detailed look at earthing arrangements see Section 6 of this guide.

This section also refers to Appendix 3 of the IET Wiring Regulations, which provides information on the characteristics of various types of protective devices.

Before you get started it is important that you assess the existing installation to determine whether or not it is capable of handling the new load or if it is suitable for alteration. If your installation is in a college or training centre workshop, for example, the existing supply will have already been assessed for safety by the lecturer or workshop technician. However, it is always advisable to check.

8.2.1 Additions and alterations to an installation

Regulation 132.16 of the IET Wiring Regulations states that:

No addition or alteration, temporary or permanent, shall be made to an existing installation unless it has been ascertained that the rating and the condition of any existing equipment, including that of the distributor, will be adequate for the altered circumstances. Furthermore, the earthing and bonding arrangements, if necessary for the protective measure applied for the safety of the addition or alteration, shall also be adequate.

This will involve a thorough visual examination of the existing installation as well as various tests to identify whether or not the installation is safe for continued service. An examination of the supplier's and the distributor's equipment, such as the meter, cut-out fuse, earthing arrangement etc., will also be necessary.

One common problem that electricians may come across in their career is where an older consumer unit fitted with circuit-breakers (see BS EN 60898) or rewirable fuses (BS 3036) but no RCDs, is replaced with an up-to-date consumer unit. A circuit-breaker or a fuse will not detect a fault between neutral and earth and yet will function without any evident problem, whereas a consumer unit with RCD protection will identify this type of fault and will not allow the circuit to energise. This might mean that a neutral to earth fault on a circuit within the installation will not have been picked up by the old consumer unit. If the electrician has not already performed an insulation resistance test on all circuits they may miss the neutral to earth fault once the consumer unit is replaced with the up-to-date version. This can lead to many complications that could have been avoided, and may increase the cost of the work being carried out as the fault will need to be identified, located and rectified before the installation can be energised – and nobody likes unexpected cost.

8.2.2 Consumer unit arrangements

In January 2015, Amendment No. 3 to the IET Wiring Regulations was published. One of the most significant changes to the regulations is the requirement that all new consumer units installed in domestic properties shall have their enclosures fabricated from a non-combustible material or be enclosed in a cabinet or enclosure fabricated from a non-combustible material complying with Regulation 132.12 (Regulation 421.1.201). How a consumer unit is arranged can have a significant impact on the safety of the user. For example, since the introduction of RCDs as a requirement for some electrical installations, people have been better protected from electric shock, and properties and equipment from damage arising from faults. One problem, however, is that when a single RCD has been used to protect an entire installation and a fault occurs, the RCD disconnects the supply from the entire installation. When this happens at night, for example, it creates a new hazard: it may leave the consumer with no means of illuminating their property. For those who may not be familiar with the location of their consumer unit, being left in darkness could present a serious hazard. It is therefore recommended that a consumer unit be installed in a such a way that, for example, the upstairs lights and downstairs sockets will be protected by one RCD and the downstairs lights and upstairs sockets by another. This means that if one of the RCDs trip then the customer will still have power to either the lighting circuit or to the final socket-outlet circuit. This will make it very unlikely that the entire building will be left without power and that a means of illuminating either of the two floors, whether via the ceiling/wall lights or via a table lamp, would still be available.

In July 2018 the 18th edition of the IET Wiring Regulations was published. One of the most significant changes with regard to requirements for the installation of consumer units and distribution boards is that the individual components within the assembly must be verified as being compatible with each other. Regulation 536.4.203 has been introduced to address known safety issues relating to devices failing, sometime catastrophically, when installed together in the same assembly with other non-compatible devices. Installers need to be aware that when selecting devices for consumer units and distribution boards that their compatibility has been verified. If an installer uses various devices not specified by the manufacturer(s) as being compatible then they (the installer) will become the assembly manufacturer and adopt the associated responsibilities.

▼ **Figure 8.1** Recommended arrangement for an RCD protecting a domestic installation

8.2.3 Heights of switches, socket-outlets and consumer units

When planning any new installations, switches and socket-outlets must be easily accessible to all persons and should therefore be installed within the required heights. To comply with building regulations, light switches and socket-outlets must be installed between 450 mm and 1,200 mm from the finished floor level. Consumer units are required to be installed between 1,350 mm and 1,450 mm above the finished floor level. If you are working on an older installation that does not meet these requirements it is not necessary to re-position all switches and socket-outlets and the consumer unit to these heights.

▼ **Figure 8.2** Required heights for switches, socket-outlets, etc.

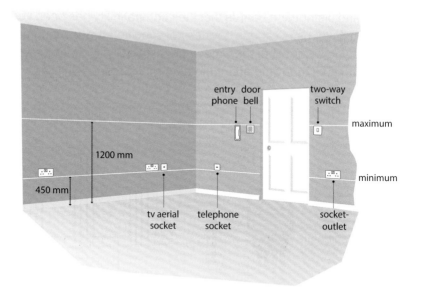

With any new installation it should be determined if additional protection is needed by means of an RCD (see Regulation 411.3.3).

When installing cables, Chapter 52 and in particular Section 522 *Selection and erection of wiring systems in relation to external influences* should be considered. These regulations cover the requirements for protecting cable against damage when they are installed within walls and partitions. There may be factors to take into consideration such as metal parts that may become live if the cable is damaged and live conductors become exposed.

8.2.4 Cable routes

You know where to mount your accessories – now you need to consider how you will get the cable to them. You could save time and money if you could simply run them in straight lines to where you need them. But you need to consider what someone might think when putting up a picture frame. In most cases, they would not install them above or below a switch or socket-outlet, however, you never know! It would be expected that they would consider the possibility that there may be a cable buried in the wall directly above, below or to either side of the equipment. So, in an ideal world, they would choose a position outside of these areas to hang the family portrait.

Figure 8.3 shows the zones where cable routes are permitted providing they have a minimum degree of protection. Full details of the minimum degrees of protection required can be found in Section 7 of the *On-Site Guide*.

8

▼ **Figure 8.3** Permitted cable routes

The aspects we have discussed are some of the fundamental ones that should be considered prior to installing any electrical installation. There are many, many more that you will become familiar with as you gain more experience studying and working in the electrical industry, for example, when drilling joists to run cables through, how far away from the top and bottom of the joists the holes need to be and, when chasing walls, the maximum depth they should be. If it is not done correctly there may be serious consequences, such as the house falling down! Keeping this in mind and always asking a supervisor will help to avoid making mistakes.

During the development of the 3rd amendment of the 17th edition of the Wiring Regulations a requirement to provide adequate support for wiring systems in escape routes so as to prevent their premature collapse in the event of a fire was introduced. This requirement has been developed further to cover all wiring systems whether or not they are in escape routes or not (521.10.202). this requirement has been introduced as a result of known safety issues that have resulted in the fatality of several firefighters.

8.3 Luminaires and lighting installations

8.3.1 Regulations and guidance

Section 559 of the IET Wiring Regulations provides the requirements for designing and installing a lighting circuit. Before you can start work on such an installation, you must find out what the customer would like. There are a number of questions that will need to be answered, such as:

(a) How many luminaires are to be installed?

(b) What kind of lamps will be used in the light fittings? For example, LED, fluorescent, halogen, incandescent etc.

(c) Are they to be installed indoors?

(d) Are they to be installed outdoors? Section 714 of the Wiring Regulations provides the requirements for outdoor lighting installations.

(e) What are the manufacturer's requirements for the luminaires?

(f) Will they all be switched on and used at the same time for long periods of time, as in an office building or a factory? (See Section 12 Diversity.)

(g) Will they be in a domestic property and not in continuous use? (See Section 12 Diversity.)

(h) What type of installation method will be used? (See table 4A2 of Appendix 4 of the IET Wiring Regulations.)

(i) What type of switches are required i.e. one-way, two-way or intermediate?

(j) Does the installation comply with the relevant parts of the Building Regulations?

These are important factors to keep in mind as they can have a significant effect on the design and cost of the installation.

At this point, if any of the customer requests do not comply with the IET Wiring Regulations the electrician must advise accordingly. For example, if a customer supplies a light fitting that does not comply with the relevant product standard, such as a fitting purchased from overseas with no manufacturer's documentation, and requests that it be installed, the installer must advise the customer that the fitting does not comply with Regulation 559.3.1 and the light fitting should not be installed.

8.3.2 Selection of equipment

Protective device

The first part of a lighting system is the consumer unit, which has been described earlier in 8.2.2.

The consumer unit will have an appropriate protective device for the lighting circuit. The rating of this protective device will have been determined when calculating cable sizes. When lighting circuits are installed in a domestic (household) premises they will be provided with additional protection by means of an RCD with a residual operating current not exceeding 30 mA as required by Regulation 411.3.4.

▼ **Figure 8.4** Protective device

Luminaires

The type and characteristic of a luminaire will vary depending upon the environment in which it is to be installed. For example, if the lighting is required in a swimming pool it must have the required IP rating to be submerged underwater. If it is being installed in a workshop the ability for the fitting to withstand mechanical impact should be considered.

Ceiling rose

▼ **Figure 8.5** Terminals at the ceiling rose

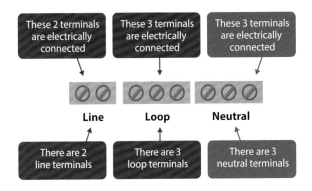

Power to a lighting circuit is usually supplied by two different methods. These are commonly referred to as the two-plate and the three plate-methods.

The two-plate method

This method requires the supply to go straight to the switch where, in an installation that uses two core and earth, only the line conductor is switched and the neutral continues through the switch to the luminaire. It is common for the neutral conductor to be connected together at the switch using a connector that complies with the requirements of the IET Wiring Regulations. By introducing a new connection you are essentially creating a new point that could introduce a safety risk. This must be borne in mind when selecting suitable components to terminate the conductors, and you must also check that they comply with the relevant standards. The two-plate method is growing in popularity and is the preferred method when using certain types of switches that need to have a neutral in order to function correctly, such as smart switches that can be controlled remotely using smart home Apps.

8

▼ **Figure 8.6** Two-plate method

The three-plate method

This method of wiring involves the supply to the ceiling rose and a switch wire to the switch and a return switch wire back to the luminaire. All three connection points of a three-plate ceiling rose are used, for example, Line, Loop and Neutral.

▼ **Figure 8.7** Three-plate method

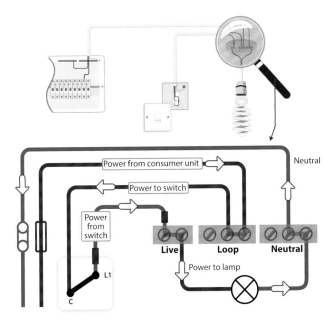

Using the three-plate method is common when more than one light needs to be operated by separate switches on a circuit. It enables the installer to provide power to multiple lighting points that can be switched individually. In the past this method has been commonly used in domestic premises for ease of installation as you can run a single cable to supply all light fittings and then just one cable to the switch from each point. The primary supply for the circuit comes into the ceiling rose and can be looped to another ceiling rose, and as many more as needed, provided that the current carrying capacity of the cable is not exceeded, the permitted volt drop is not exceeded and disconnection times can be achieved. This is shown in Figure 8.8.

▼ **Figure 8.8** Looping power

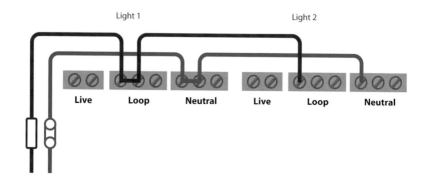

Cable

The line conductor can be either a single-core cable or contained within a multi-core cable. The size and type of the cable will be determined using a cable calculation that takes into account the overall load of the luminaires required as well as the environment into which it is to be installed.

The cpc is not shown in the following diagrams as these diagrams simply represent the functionality of the circuit. However, when installing any electrical system it should be determined what is needed in the way of a protective conductor (see Section 6 Earthing and Bonding).

Switching devices

Depending on the requirements of the system there are various switching devices that can be used. It is important to consider the type of fittings that are going to be installed and refer to manufacturer's instructions so that the correct switches can be selected for the installation. Some lamps and their controlgear draw a lot of current when they are first switched on; this needs to be considered when selecting the switching device. The amount of current that switches are designed to use can vary. If an underrated switch is installed which cannot handle the current, there is a risk of the switch being damaged and can potentially overheat and lead to a fire.

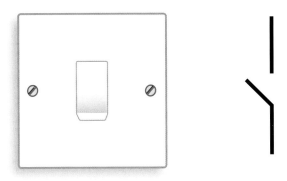

One-way switch

If the light needs to be switched on and off from only one place then a one-way switch, like the one below, will be required.

▼ **Figure 8.9** One-way light switch

The following image shows a simple one-way switch wiring configuration. This is a typical method used that requires a light to be controlled from only one position.

8

▼ **Figure 8.10** Connection of conductors in a one-way light switch

Two-way switch

If a light needs to be switched on and off from two separate locations, a two-way light switch will be required. The configuration of a two-way light switch is slightly different, as you can see below. The C (common) will link to either L1 or L2. L1 and L2 are never linked together on a two-way switch. Some switches may have their terminals marked with L1, L2, L3 as the manufacturer chooses to replace the C with L1 and L1 with L2 and L2 with L3.

▼ **Figure 8.11** Two-way light switch

The following image shows a two-way conventional wiring configuration. This method of connecting a lighting circuit is commonly used in commercial or industrial environments using single-core cable.

▼ **Figure 8.12** Two-way conventional

The following image shows a two-way conversion wiring configuration. This method of connecting a lighting circuit is commonly used in domestic properties using flat twin and earth and flat three-core and earth cable. This method is favoured as it only requires an additional 3-core cable to be installed between the switches.

▼ **Figure 8.13** Two-way conversion

Intermediate switch

If the lighting system needs to be switched on and off from three or more locations, such as in a stairwell rising over several floors, an intermediate switch will be required. Looking at the figure below you can see how the terminals are connected to one another during operation of the switch.

▼ **Figure 8.14** Intermediate light switch

The following image shows an intermediate conventional wiring configuration commonly used in commercial and industrial installations.

▼ **Figure 8.15** Intermediate (conventional) wiring configuration

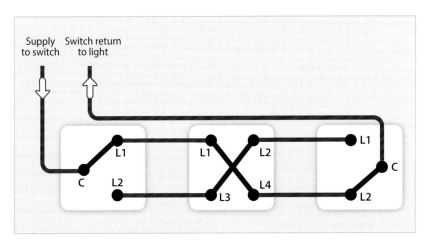

The following image shows an intermediate conversion wiring configuration that is commonly found in domestic installations.

▼ **Figure 8.16** Intermediate (conversion) wiring configuration

8.3.3 Installation conditions

There are not many places in the developed world where you won't find lighting of some sort; from ambient lighting for effect (such as pond lighting) to functional lighting for illuminating large areas for a workforce to operate safely (such as portable lighting for highway maintenance).

Correct selection of lighting and cable

When selecting the appropriate lighting for a particular task, the condition of the environment should be taken into consideration, not only for the luminaire and the light it can provide but also for the entire wiring system. In Section 1 of this Guide there is a list of the cables that can be used for wiring systems. Some are suitable for small domestic installations where there will be little risk of heavy mechanical impact. Others have varying grades of protection, from steel braided wire to complete metal sheathing, such as SWA and MIMS, that will withstand significant mechanical impact. Of course, these vary dramatically in price and the time it takes to install. Usually more skill and time is required for the installation of cable that has increased resistance to mechanical damage.

The light fitting to be installed should also be chosen to suit the environment. To best select a light fitting, an assessment of the environmental conditions should be made – not only of how it is during installation, but also what the area will be used for after the installation is complete.

For example, if you are installing a lighting circuit in a sauna you would want a wiring system that can withstand the moisture and high temperatures expected in that environment.

Access

Lighting is generally installed at high level, so some form of access equipment, such as a stepladder, will be needed. More complicated access equipment, such as a tower or scaffolding, can add considerable cost and time to the installation and customers should be informed of this prior to work commencing.

> **Remember ...**
>
> When connecting conductors into terminals, use a wiring or circuit diagram and focus on one conductor at a time. When working with two-way switches it is good practice to connect your strappers first; L1 at one switch should be connected to L1 at the other switch and L2 at one switch should be connected to L2 at the other switch (when flat twin and earth cable is used for switch wires the line conductor needs to be identified with brown sleeving). And remember there should never be a neutral conductor connected to a standard a one-way, two-way or intermediate switch for lighting. As mentioned earlier there are switches that require a neutral to function but these are generally for lighting control gear or SMART installations.

8.4 Radial final circuits (power)

8.4.1 Regulations and guidance

Radial final circuits for socket-outlets are not as common in the UK as ring final circuits. However, some electricians prefer to use radial final circuits to supply socket-outlets. Appendix 15 of the IET Wiring Regulations provides guidance and information on the installation of radial final circuits used for socket-outlets.

A radial circuit is a simple circuit to install. A 230 V radial circuit will consist of two live conductors (line and neutral) and a cpc. These will be fed from an appropriate protective device, such as a circuit-breaker or fuse and RCD.

8.4.2 Selection of equipment

The circuit will start at the consumer unit. This is where a suitable protective device will be installed. In new installations, this will likely be a circuit-breaker (see BS EN 60898). The cable used will be sufficient to handle the expected current that will be flowing through the circuit. Guidance from Table H2.1 of the *On-Site Guide* can be used. However, it should be borne in mind that installation conditions may affect the current-carrying capacity of the cable. The socket-outlets should be manufactured to BS 1363. Regulation 553.1.3 and Table 55.1 from the IET Wiring Regulations provide more information on the minimum requirements for plugs and socket-outlets for low voltage circuits. There are now socket-outlets available that have USB circuits incorporated into them. BS 1363 states that USB circuits incorporated in a socket-outlet shall conform to the relevant standards for information and technology equipment.

8.4.3 Maximum number of socket-outlets

Although there is no specified maximum number of socket-outlets that can be installed on a final radial circuit, the requirements of Regulation 433.1.204 (explained in more detail in Section 8.7 of this Guide) shall be met.

Remember: if a cable passes through or runs alongside thermal insulation it significantly reduces the current-carrying capacity of the cable (see Section 7 Cable Calculations).

8.4.4 Minimum and maximum heights for socket-outlets and switches

To comply with the Building Regulations, socket-outlets in new installations must be installed above 450 mm and below 1200 mm. This is to ensure equality of access – so that people with limited reach, such as wheelchair users, can still access switches and socket-outlets. Regulation 553.1.6 requires socket outlets to be positioned in locations to allow plenty of clearance from worktops, shelves and other objects or furnishings so that damage to the flexible cable and the plug of any equipment when being plugged in or unplugged is prevented.

8.4.5 Maximum area that a radial final circuit can serve

A 30/32 A radial final circuit using 4 mm^2 copper conductor thermoplastic or thermosetting insulated cables should serve an area not exceeding 75 m^2. For a 20 A radial final circuit using 2.5 mm^2 copper conductor, thermoplastic or thermosetting insulated cables should serve an area not exceeding 50 m^2.

Table H2.1 in the *On-Site Guide* provides information on the required CSA of conductors and the maximum area that a radial final circuit can serve, as well as the rating of the protective device.

8.5 Ring final circuits

The definition of a ring final circuit (from Part 2 of the IET Wiring Regulations) is:

A final circuit arranged in the form of a ring and connected to a single point of supply.

8.5.1 Regulations and guidance

Ring final circuits are used to supply accessories such as socket-outlets complying with BS 1363-2 and fused connection units complying with BS 1363-4, and are almost unique to the UK. In 1942 the Electrical Installations Committee determined that the ring final circuit offered a more efficient and cheaper wiring system that would support a greater number of socket-outlets. Section 553.1 of the IET Wiring Regulations covers the requirements for socket-outlets installed on a ring final circuit.

Regulation 433.1.204 provides details of the type and rating of the protective device that must be selected, the minimum CSA of the line conductors, and the cpc, including the minimum cpc. Appendix 15 of the IET Wiring Regulations and Appendix H of the *On-Site Guide* provide guidance on ring final circuits.

The ring final circuit is intended to evenly distribute the load on the circuit so that the maximum current-carrying capacity of the cable is not exceeded for long periods of time.

8

The illustration below shows how the load is distributed evenly over two legs of the ring (the current flow is shown with the red arrows). This reduces the strain on the cable and the risk of it being overloaded.

▼ **Figure 8.17** Even distribution of load over a ring final circuit

8.5.2 The required CSAs for conductors on a ring final circuit

The CSA of the live conductors (line and neutral) for a ring final circuit should be 2.5 mm^2 and the cpc should be no less than 1.5 mm^2.

8.5.3 Maximum number of socket-outlets

An adequate number of socket-outlets should be provided so that equipment can be supplied from a conveniently accessible socket-outlet, taking into account the length of flexible cable normally fitted to portable appliances (see Regulation 553.1.7). This basically means that you can have (within reason) as many socket-outlets as required for the installation. The IET Wiring Regulations do not state a maximum number of socket-outlets that can be installed on a ring final circuit. However, Regulation 433.1.204 clearly states that, given the intended condition of use, it should be unlikely to exceed for long periods of time the current-carrying capacity (I_z) of the cable. Consequently, the responsibility sits with the designer to determine the number of socket-outlets to be installed.

8.5.4 Maximum area that a ring final circuit can serve

A ring final circuit should serve an area not exceeding 100 m². Table H2.1 in the *On-Site Guide* provides information on the required CSA of conductors and the maximum area that a ring final circuit can serve, as well as the rating of the protective device and the minimum CSA of the cpc. This table can also be referred to when looking at radial circuits.

▼ **Figure 8.18** Table H2.1 from the *On-Site Guide*

Type of Circuit	Overcurrent protective device rating (A)	Minimum live conductor cross-sectional area* (mm²)		Maximum floor area served (m²)
		Copper conductor thermoplastic or thermosetting insulated cables	Copper conductor mineral insulated cables	
1	2	3	4	5	6
A1	Ring	30 or 32	2.5	1.5	100

8.6 Spurs

8.6.1 What is a spur?

A spur is simply an individual cable coming from a point on a circuit to supply an accessory or piece of equipment. This can be a socket-outlet or supply to equipment. There are two types of spurs: unfused and fused. Both are added to a final circuit by connecting a cable into an existing socket-outlet or joint box and then to the new socket-outlet or fused connection unit.

8.6.2 Adding a spur to a circuit

A spur is an amendment or addition to an existing circuit that is permanently connected directly to a ring final circuit or radial circuit.

When adding one single or double socket-outlet to a final socket-outlet circuit there is no requirement to install a fused connection unit providing that only one single, double or multiple socket-outlet is connected to a single point of the circuit. This is to prevent any single branch becoming overloaded.

If more than one spur is required to be installed from a single point on the ring final or radial final circuit, additional protection is required by means of a fused connection unit manufactured to BS 1363-4, as shown in Figure 8.19. Providing the requirements of Regulation 433.1.204 are met, as many socket-outlets or permanently connected equipment can be fed from the existing circuit.

▼ **Figure 8.19** Final ring circuit with spurs

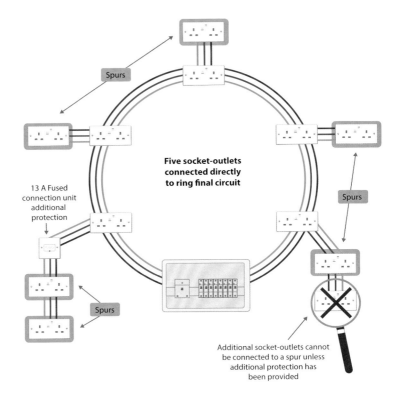

▼ **Figure 8.20** Connection of cpc to a socket-outlet including earth tail

By ensuring that the cpc have been terminated correctly the general safety of the circuit will be improved, especially under fault conditions.

8.7 Shower circuit

One of the simplest circuits that you will ever have to install is that which only serves one piece of equipment, such as a shower. Although there may be processes that are more complicated going on inside the equipment, the supply to the equipment is straightforward.

8.7.1 A circuit supplying a shower

Each circuit supplying a shower will require:

(a) appropriately selected and rated protective devices for overload, short circuit and earth fault protection;
(b) cable with suitable current-carrying capacity;
(c) an isolator; and
(d) the equipment itself, i.e. the electric shower.

A circuit for an electric shower shall be protected by an RCD in accordance with Regulation 701.411.3.3 of the IET Wiring Regulations with the characteristics specified in Regulation 415.1.1

and be designed to supply the full load. Diversity shall not be applied on this type of circuit (see Section 12 of this Guide).

8.8 Separated Extra-Low Voltage systems (SELV)

For the regulations related to SELV, see Section 414 of the IET Wiring Regulations.

▼ **Figure 8.21** SELV Transformer

8.8.1 BS EN 61558-2-6 safety isolating transformer

For circuits that require further protection so that the risk of electric shock is reduced, for example, extractor fans installed within a location containing a bath or shower, a SELV supply can be used.

This type of system separates incoming supply to the property from the supply to the equipment by using an isolating transformer. This means that this type of transformer not only reduces the voltage to supply the equipment but also completely separates the electrical supply from the equipment. You can see in Figure 8.21 that there is no physical electrical connection between the conductors. The power that is supplied to the equipment is produced by mutual inductance (you will learn more about this process throughout you electrical course).

8.8.2　How does it work?

SELV circuits use a transformer to separate (there is no physical connection between the conductors of the primary and the secondary sides) the electrical supply from the circuit and reduce the voltage to a safe level. This type of transformer can be used to supply electrical equipment within the permitted zones in bathrooms.

As well as the voltage being reduced, the circuit is electrically separated from the supply to the property and the general mass of Earth, so the risk of electric shock from a single fault on the equipment is greatly reduced. To understand this in more detail refer to Section 6 Earthing and Bonding.

8.9 Protective Extra-Low Voltage systems (PELV)

For the regulations related to PELV, see Section 414 of the IET Wiring Regulations.

▼ **Figure 8.22** PELV Transformer

8.9.1 BS EN 61558-2-6 safety isolating transformer

An isolating transformer that has all the characteristics of SELV except that it does not provide electrical separation from earth. This type of circuit should be considered when the characteristics of SELV cannot be achieved and maintained throughout the life of the installation. A typical example of where you would find a PELV transformer on a circuit is in a computer.

As you can see from the image above, the secondary winding of the transformer has its neutral connected to earth.

8.10 Functional Extra-Low Voltage systems (FELV)

For the regulations related to FELV, see Section 411.7 of the IET Wiring Regulations.

▼ **Figure 8.23** FELV Transformer

A transformer that has the characteristics of FELV uses extra-low voltage. However, it is not adequately protected from accidental contact with higher voltages in other parts of the circuit. FELV is used where extra low-voltage is required for functional purposes, such as machine control systems.

	Test your Knowledge
1.	What are the two different types of spurs?
2.	Using Table H2.1 from the *On-Site Guide* (shown earlier in this section), what are the two overcurrent protective device ratings that can be used for a ring final circuit?
3.	Three additional double socket-outlets are to be installed on one branch of a spur from a ring final circuit. What is required, with regards to overload protection, so that the additions meet the requirements of the IET Wiring Regulations?
	For the below questions see Section 559 of the IET Wiring Regulations.
4.	What is the maximum permitted operating voltage of a lighting circuit that includes a ceiling rose or lamp holder?
5.	Which regulation deals with stroboscopic effect?

8

6.	Have a look at the following diagrams and state whether the light will be on or off.
7.	Describe two characteristics of SELV circuits that protect the user from electric shock?
8.	What is the difference between a SELV transformer and a PELV transformer?
9.	What section of Part 7 of the IET Wiring Regulations should be referred to when installing a shower?
10.	Where might you find FELV transformers?

Inspection and Testing 9

After this section, you will have awareness of the following topics:

- ▶ The reason for testing
- ▶ Initial inspection
- ▶ Testing procedures

9.1 Why test?

The reason behind inspecting and testing an electrical installation is to find out whether a new installation is safe to be energised and handed over to the client, or whether an existing installation is safe to remain in service until the next inspection is due. The inspector must verify that the existing installation, including the equipment that belongs to the DNO or the meter operator, is adequate and safe. If any of the existing equipment does not support the planned new installation or is in poor condition, the installer should not proceed.

Note: The term 'inspector and tester' has been used throughout this section; this is usually, but not necessarily, the same person.

9.1.1 Legal aspects

The persons carrying out the inspection and testing must comply with the Electricity at Work Regulations 1989 (EWR). The EWR requires that whoever does the work has to be competent to do so. When testing electrical installations, some of the tests are carried out on live parts, so, it is extremely important that the tester is competent to carry out this work. Installations that comply with BS 7671 will, in most circumstances, meet the requirements of the EWR. Note that electrical contractors that issue inaccurate or false electrical installation certificates may be liable to prosecution.

9.1.2 Responsibilities of the inspector and tester

The inspector and tester must:

- (a) be competent to do the work, as required by EWR, Regulation 16;
- (b) be familiar with the requirements of the most recent version of the IET Wiring Regulations;
- (c) have a good knowledge of the type of installation that they are inspecting;
- (d) understand the inspection and testing procedures; and
- (e) use testing equipment manufactured to the required standards and in accordance with the HSE Guidance Note GS38 *Electrical test equipment for use by electricians*.

It is the responsibility of the inspector and tester:

- (a) to identify that reasonable measures are in place to reduce the risk of injury to persons or livestock;
- (b) to identify that reasonable consideration has been given to installation methods so that the risk of damage to property as a result of an electrical installation is reduced;
- (c) to compare the inspection and testing results with the design specifications; and
- (d) to make a judgement on the condition of the installation and advise on remedial works to the client.

Where the inspection highlights an extremely dangerous situation, for example, persons or livestock are at immediate risk of electric shock or an imminent fire hazard is evident, urgent action is advised to remove the danger. As the expert, the inspector and tester has been employed to identify electrical problems and, therefore, is advised to exercise judgement to secure the area and inform the client immediately, followed up in writing. Once permission has been obtained, the danger should be removed by isolating the circuit or circuits in question.

9.1.3 Why is verification important?

The reason initial verification is carried out on an electrical installation is to confirm, by way of visual and physical inspection as well as testing, during construction and on completion, that the installation complies with the design and construction aspects of the IET Wiring Regulations as well as the relevant parts of the Building Regulations.

While carrying out the initial verification you will be looking for non-compliant installation methods as well as non-compliant equipment and materials. This can be anything from an incorrectly terminated protective equipotential bonding conductor (see Section 6 of this Guide) to equipment from overseas that does not meet the requirements of EU and British Standards.

The following checks should be carried out before testing:

(a) leads of test equipment are in good condition and fit for purpose;
(b) test instruments are to be calibrated with a valid, current calibration certificate or proof of an ongoing accuracy check;
(c) probes should have no more than 4 mm of exposed tip (GS38 recommends 2 mm);
(d) test equipment in good repair (remember that test equipment can be dropped and damaged even though it is within its calibration period);
(e) all necessary PPE is in good condition;
(f) other people who may be affected by the testing are informed before the testing is carried out; and
(g) all electrical equipment is disconnected from the circuits that are being tested, including lamps.

9.2 Initial inspection

9.2.1 Why do you need an inspection?

Regulation 641.1 of the IET Wiring Regulations states that:

Every installation shall, during erection and on completion before being put into service, be inspected and tested to verify, so far as is reasonably practicable, that the requirements of the Regulations have been met.

9

Precautions shall be taken to avoid danger to persons and livestock, and to avoid damage to property and installed equipment, during inspection and testing.

The IET Wiring Regulations requires that the following information about the installation that is to be tested is made available to the person carrying out the testing procedures:

- **(a)** maximum demand.
- **(b)** number and type of live conductors (some commercial and most larger industrial premises have a three-phase supply, which consists of three or four live conductors. Smaller domestic premises usually have two, i.e. line and neutral).
- **(c)** type of earthing arrangement (see Section 6 of this Guide).
- **(d)** nominal voltage (in most supplies in the UK, 230 V or 400 V with an allowable tolerance, see Section 7 of this Guide).
- **(e)** load current and supply frequency (see Section 3 of this Guide for frequency).
- **(f)** prospective fault current at the origin (this is the highest expected short circuit or earth fault current that can be expected at the incoming supply of the installation).
- **(g)** external earth fault loop impedance (EFLI, with the measurement value of Z_e): this is the resistance of the line and earth conductor from the owner's installation to the distribution transformer.
- **(h)** type and rating of overcurrent device.

9.2.2 Sequence of tests

The tests to be carried out, where relevant, during the initial verification of new installations are listed in Section 9 of the *On-Site Guide*.

Testing and inspection assessments carried out by training providers are considered practical assessments and the candidate will therefore be required to wear appropriate PPE. In most cases, safety boots that comply with the relevant standard, are the minimum requirement.

9.3 The tests in detail

▼ **Table 9.1** Tests for initial verification

Test	Corresponding Regulation
Tests to be carried out before the power supply is connected (i.e. isolated)	
Continuity of protective conductors, including continuity of protective equipotential bonding (see 9.4.1) and continuity of circuit protective conductors (see 9.4.2).	Regulation 643.2.1
Continuity of ring final conductors (see 9.4.3).	Regulation 643.2.1
Insulation resistance (see 9.4.4).	Regulation 643.3
Polarity (by continuity method) (see 9.4.5).	Regulation 643.6
Tests to be carried out after the power supply has been connected (energised)	
Check polarity of supply using an approved voltage or polarity tester (see 9.5.1).	GS38
External earth fault loop impedance (Z_e) (see 9.5.2) and earth fault loop impedance (Z_s) (see 9.5.3).	Regulation 643.7.3
Prospective fault current (see 9.5.4).	Regulation 643.7.3.201
Phase sequence (see 9.5.5)	
Functional testing of equipment including RCDs (see 9.5.6).	Regulation 643.10

Some things you will need:

(a) appropriate PPE;

(b) insulated screwdriver set including torque screwdriver suitable for the application;

(c) suitable test equipment that complies with the guidance given in GS38, including;
   (i) low resistance ohms or continuity tester;
   (ii) insulation resistance tester;
   (iii) earth fault loop impedance tester;
   (iv) RCD tester;
   (v) earth electrode tester; and
   (vi) a polarity tester for three phase power supplies.
(d) a lead to link conductors at the consumer unit or some other means of connecting cable together;
(e) a connector block; and
(f) a wandering lead (sometimes called a long lead).

You will also need to consider how you are going to deduct the resistance of the test leads from your final result. The reason you need to do this is that you will be looking to identify the value of the circuit; the test leads are not part of the circuit and therefore their resistance should not be included in the final result. Reducing the resistance can be done by either:

(a) measuring the value of the resistance by holding the test leads together and making a record so that it can be deducted from the end result; or
(b) as most test instruments have a zero or null function (similar to the way you would zero a set of kitchen scales), carry out the procedure as instructed by the manufacturer's guidance.

▼ **Figure 9.1** Nulling/zeroing the tester

The following sequence of tests is to be carried out before the electrical supply is connected.

9.4 Dead tests

9.4.1 Continuity of protective equipotential bonding conductors

This test is to identify that the protective equipotential bonding conductors are continuous or unbroken and, subsequently, correctly connected.

The purpose of this test is to ensure that all exposed- and extraneous-conductive-parts will be at the same potential, hence the word 'equipotential' ('equal' + 'potential'). In order to achieve this, it is recommended that the resistance of the bonding conductors should not exceed 0.05 Ω.

For this test, you will need test equipment (low resistance ohms meter) that complies with the guidance given in GS38.

To complete this test you need to follow the simple steps in order.

ONE: isolate the supply (see the safe isolation procedure in Section 2 of this Guide and the Student's Guide video series on IET.TV).

TWO: disconnect one end of the conductor at the consumer unit.

THREE: measure the resistance of the test leads or, better still, null/set them to zero.

FOUR: connect one test lead to the disconnected conductor at the consumer unit.

FIVE: connect the other lead to the opposite end of the conductor (depending on the distance between the ends of the bonding conductor you may be required to use a wandering lead).

SIX: if you haven't nulled the test instrument, remember to subtract the resistance of the leads from the total resistance shown on the meter. This will give you the resistance of the bonding conductor.

SEVEN: ensure that the bonding conductor is reconnected on completion of the test (if this is not done the installation will not be bonded).

Whilst carrying out this test, a visual inspection can also be done to ensure that the correct type of earthing clamp is present, complete with the correct label, and that the conductor is one continuous length (has not been broken and re-connected) if it is bonding more than one service or extraneous conductive part.

9.4.2 Continuity of cpc

For reasons of safety, the cpc is one of the most important conductors in the installation. If something goes wrong at any point, such as a line to earth fault, the cpc will carry sufficient fault current around the earth fault loop impedance path in order to operate the protective device.

As the person doing the tests, you will need to verify that the cpc of the radial circuits are in place, and are connected throughout the circuits that you are testing in the installation.

The instrument that you will use to carry out this test is the low resistance ohm meter (sometimes called a *continuity tester*) and should be set on the lowest scale for measurement.

This test is commonly referred to as a dead test, which means that it must be carried out on an isolated circuit. Continuity of cpc can be determined using one of two methods. When carrying out either of these tests on a circuit with light switches, for example, it will be necessary to close the switches in order to test parts all parts of the circuit to obtain an accurate measurement.

Method 1 (also known as the link method)

You will see that on an IET model form for schedule of inspections there are two permissible values that may be recorded when verifying the continuity of the cpc. These are $R_1 + R_2$, or just R_2. R_1 represents the line conductor and R_2 represents the cpc. Method 1 essentially uses the line conductor as an extended test lead. It runs alongside the cpc so it might as well be used to help obtain the continuity reading for the cpc.

For this test, you will need:

(a) test equipment (low resistance ohm meter) that complies with the guidance given in GS38;
(b) appropriate PPE; and
(c) a temporary link, a short length of cable or a connector so that the live and the cpc can be joined together at one end of the circuit, usually at the consumer unit.

To complete this test you need to follow these steps in order.

ONE: the supply will be isolated (see Section 2 of this Guide for information about the safe isolation procedure).

TWO: measure the resistance of the test leads or, better still, null/set the tester to zero.

THREE: using a short length of cable or a connector block, connect the line and the cpc together at the consumer unit.

FOUR: the test instrument should be set to a low resistance setting and the probes should then be applied to the cpc and line conductor at each accessible point of the circuit, i.e. at the switch and ceiling rose. During this test, current will flow from one end of the test instrument, for example, from the red lead, through the line conductor of the circuit to the consumer unit then through the temporary link back through the cpc to the test instrument, at which point the resistance will be measured and displayed on the test instrument.

▼ **Figure 9.2** Example of Method 1 on a radial final circuit

▼ **Figure 9.3** Example of Method 1 on a lighting circuit

FIVE: the highest reading obtained for each circuit should be recorded onto the Schedule of Test results. The readings should increase as you move further away from the consumer unit.

The result obtained using this test method can be added to the external earth fault loop impedance to calculate the total resistance of the earth fault loop impedance path (measurement value Z_s).

Method 2 (also known as long lead or wandering lead)

This method will prove that the cpc is continuous. This method is usually only used where the circuit is wired in steel enclosures (trunking/conduit/ducting) and parallel paths to the cpc may be present. It is also useful if there are many enclosed metal fittings and dismantling them would be impractical.

You will need:

(a) a low ohm resistance test meter that complies with the guidance given in GS38;
(b) appropriate PPE; and
(c) a wandering lead (wandering leads are available to purchase at most electrical wholesalers).

9

There is no need to carry out this testing method if Method 1 has already been carried out.

To complete this test you will need to follow these simple steps in order.

ONE: isolate the supply (see Section 2 of this Guide for information about the safe isolation procedure).

TWO: measure the resistance of the test leads or, better still, make them zero.

THREE: either connect the wandering lead directly to the test instrument or connect the lead from the test instrument to one end of the wandering lead. The wandering lead from the test instrument can be connected to the earth terminal at the consumer unit. Measurements shall be taken from all accessible points on the circuit. From all the resistance values that have been measured, the highest reading must be recorded and written in the Schedule of Test results.

▼ **Figure 9.4** Example of Method 2 on a radial circuit

▼ **Figure 9.5** Example of Method 2 on a lighting circuit

FOUR: you can then take the tester and the short lead, hold the test probe against the earth terminals at each accessible piece of equipment and accessory at it's point of connection to the circuit you are testing ensuring that you have identified the furthest point on the circuit, and test it.

FIVE: the highest reading is to be recorded on the Schedule of Test results under (R1 + R2) or under R2.

9.4.3 Continuity of ring final circuit conductors including protective conductors

The purpose of this test is to confirm that:

(a) the conductors form a complete ring. When conductors are not connected to complete a full ring, various parts of the circuit may have more current passing through it than the cable is designed to handle. This may lead to overheating and fire.

(b) there are no interconnections. Interconnections suggest that the circuit may have been 'nected in such a way that there are bridging cables between two parts of the circuit. This can create additional strain on certain parts of the circuit and can also make it difficult to verify other test results.

(c) the polarity is correct in all socket-outlets. This test identifies that the line conductor is connected to the live terminal of the socket-outlet and the neutral is connected to the neutral terminal of the socket-outlet.

(d) there may be an increased risk of electric shock if the polarity is the wrong way around, for example, if a lamp that has an Edison screw type fitting is plugged into a socket with incorrect polarity.

When this test is carried out correctly it also gives you the R_1 and R_2 value of the ring final circuit and can be used to identify any spurs in the circuit.

The instrument that you will use to carry out this test is, once again, the low resistance ohm meter and, as before, should be set on the lowest value possible.

You will need:

(a) a low ohms resistance test meter that complies with the guidance given in GS38;
(b) appropriate PPE; and
(c) three connectors.

To complete this test you need to follow these simple steps in order.

ONE: isolate the supply (see section 2 of this Guide for information about the safe isolation procedure).

TWO: measure the resistance of the test leads or, better still null the test instrument.

THREE: identify each leg of the ring (incoming and outgoing); this is usually done at the installation stage by marking the cables with either **in and out** or **1 and 2** etc.

FOUR: test between the line conductors of the ring. Record this value on the Schedule of Test results.

▼ Figure 9.6 Testing the live conductors

FIVE: test between the neutral conductors of the ring. Record this value on the Schedule of Test results. This value should be the same as the line conductor.

▼ **Figure 9.7** Testing the neutral conductors

SIX: test between the ends of the cpc of the ring. Again, record this value on the Schedule of Test results. If the conductor sizes are smaller than the phase and neutrals, this value will be higher.

Figure 9.8 Testing the cpc

SEVEN: securely join together, with a connector block, the incoming line to the outgoing neutral conductors and the outgoing line to the incoming neutral conductors as shown in Figure 9.9.

Figure 9.9 Line from leg one connected to neutral from leg two

EIGHT: test between line and neutral at each socket-outlet on the circuit. The resistance should be the almost same at each socket-outlet. A higher reading should be investigated, as it may be a spur on the circuit an should be recorded on the inspection schedule or a loose connection.

NINE: disconnect the neutral conductors from the connector block and connect the cpc and Line conductors as shown in Figure 9.10. Repeat the test at the line and cpc terminals at each socket-outlet. Operate each switch on the socket-outlets to confirm polarity.

▼ **Figure 9.10** Line from leg one connected to cpc from leg two

9.4.4 Insulation resistance Regulation 643.3

Important: this test can cause damage to electrical and electronic equipment if the procedures are not carried out correctly. There is also a risk of electric shock from the DC produced by the test instrument.

This test is carried out to determine that the resistance of the insulation around the conductors is sufficient and that there are no short circuits between the live conductors or live to cpc. The insulation is designed to stop current escaping from the conductor. However, cable insulation can deteriorate with age. There are various other ways that the resistance of the insulation may weaken, for example, cables that have been crushed beneath floorboards, damaged by clips or suffered damage due to being pulled through holes in joists, etc.

This test can be carried out on a complete installation or on a single circuit, whichever is suitable or required. This test is necessary to perform in order to find out if there is likely to be any leakage of current through the insulated parts of the installation.

The instrument used to carry out this test is an insulation resistance tester. This method works a little like a pressure test. It subjects the conductors to a high voltage for a short time to ensure that the insulation surrounding them contains the voltage. If the insulation is not sufficient there is a risk of a short circuit or line to earth fault. Table 9.2 shows the required setting for the test instrument in relation to the voltage of the circuit.

▼ **Table 9.2** Minimum values of insulation resistance (Table 64 from the IET Wiring Regulations)

Nominal voltage of the circuit	DC Test voltage to be applied (V)	Minimum insulation resistance (MΩ)
0 V – 50 V (SELV and PELV)	250	0.5
Up to and including 500 V with the exception of SELV and PELV but including FELV	500	1.0
Above 500 V	1000	1.0

Although an insulation resistance value of not less than 1.0 MΩ complies with the IET Wiring Regulations (the resistance is usually much higher), where the measured insulation resistance is less than 2.0 MΩ, each circuit should be tested separately. This will help to identity, firstly, which circuit has a lower value and, secondly, a possible defect in the insulation of the conductor.

Before carrying out the test, ensure that:

    (a) all lamps are removed;
    (b) all electronic devices are removed;
    (c) all plug tops are removed from the socket-outlets;
    (d) all protective devices are in the 'on' position and that the main switch is in the off position..

To complete this test you need to follow the below steps in order.

    ONE: set the insulation resistance tester to the correct voltage needed for the circuit to be tested (500 V for 230 V circuits).

    TWO: push the test button with the test leads disconnected. The resistance shown on the screen should be the high Ω reading, for example, >999 MΩ.

THREE: join the test leads together and operate the test button again. The resistance shown on the screen should be very close to 0.0 Ω. Steps two and three identify that the test instrument is working.

FOUR: the quickest way to identify that the insulation resistance is sufficient on all circuits is to test them all at the same time. When testing lighting circuits that have two way or intermediate switching, each switch must be operated and the circuit re-tested after each change of switch position so that all conductors including strappers have been tested. The terminals that the test probes need to be applied to are shown as follows.

▼ **Figure 9.11** Testing insulation resistance between line and neutral

▼ **Figure 9.12** Testing insulation resistance between Neutral and Earth

▼ **Figure 9.13** Testing the insulation resistance between live and earth

FIVE: record the test results on the Schedule of Test results.

SIX: use Table 10.3.3 in the *On-Site Guide* to verify that the results comply with The Wiring Regulations.

The following tests are live tests and should only be carried out under the supervision of an electrically skilled person, such as a lecturer.

9.4.5 Polarity

This test is to confirm that all:

(a) protective devices are connected to the line conductors only;
(b) single-pole devices are also connected to the line conductors only;
(c) ES lamp holders have the centre pin connected to the line conductor; and
(d) accessories are correctly connected.

In most cases the polarity can be confirmed by operating the switch at the same time as the continuity of cpc test is carried out. When the switch is operated the reading from the low ohm meter will indicate a low resistance when the switch is closed and a high resistance when the switch is open. Polarity can also be checked visually by verifying that the cores have the correct colour identification at their terminals.

Using the low resistance ohm meter with the lowest value of Ω possible, complete this test by following the simple steps in order.

ONE: isolate the supply (see Section 2 of this Guide for information about the safe isolation procedure).

TWO: put the main switch in the off position, remove all fuses or put protective devices in the off position and remove all lamps and equipment.

THREE: using a short lead with a crocodile clip or a connector block on each end, connect the line and cpc of the circuits to be tested together at the consumer unit.

FOUR: measure the resistance of the test leads or, better still, null the test instrument.

FIVE: at each point of each circuit, test between the line (phase) and the cpc (either at the switch or at the lampholder).

SIX: close the switch that is controlling the light; the test instrument should read a low resistance. Opening the switch will then result in a high resistance. This confirms that the line conductor has been switched and therefore confirms polarity.

9.5 Live tests

9.5.1 Polarity

You need to test that when you switch on the supply, you have a voltage present. This test is carried out at the line and neutral terminals of the incoming supply to the consumer unit.

The instrument that you need to use must comply with GS38 and be an approved voltage indicator.

To complete this test you need to follow the steps in order.

ONE: place the test probes of the voltage indicator onto the line and neutral terminals of the incoming supply at the main switch. Your device should indicate a live supply.

TWO: place the test probes of the voltage indicator onto the line and earth terminals of the incoming supply at the main switch. Your device should indicate a live supply.

THREE: place the test probes of the voltage indicator onto the neutral and earth terminals of the incoming supply at the main switch. Your device should indicate NO supply.

9.5.2 Earth fault loop impedance (Z_e) Regulation 643.7.3

Z_e is a measurement of the external earth fault loop impedance of the installation.

To carry out this test the installation should be isolated from the supply and the main earth should be disconnected from the main earthing terminal. This is to avoid any parallel paths through any other earthed metalwork within the building.

The instrument used for this test is an earth fault loop impedance tester. The assumed values for Z_e can be found by completing the following steps in order.

ONE: acquire the necessary PPE and insulated tools.

TWO: isolate the supply to the consumer installation (the incoming supply must remain live to carry out this test).

THREE: disconnect the earthing conductor (this is essential to avoid obtaining readings from parallel earth paths).

FOUR: set your instrument to loop Z_e (you may have to refer to the manufacturer's instructions for the test instrument to find out what the correct setting is).

FIVE: if using the two-lead instrument, one lead should be connected to the disconnected incoming main earth conductor and the other lead to the incoming line conductor on the supply side of the main switch.

▼ **Figure 9.14** Earth fault loop impedance two lead test

If using the three-lead instrument, two leads should be connected to the disconnected incoming main earth conductor and the other lead to the incoming line conductor on the supply side of the main switch.

SIX: press the test button and record the result onto the Schedule of Test results (the incoming supply should remain live).

9

▼ **Figure 9.15** Earth fault loop impedance three lead test

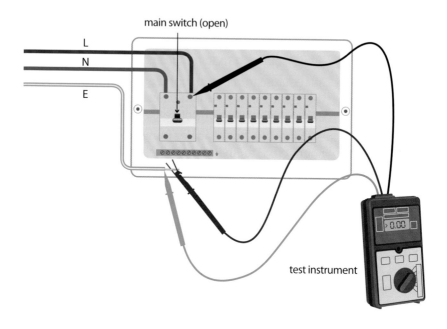

SEVEN: reconnect the incoming main earth conductor. This is a very important step. When being assessed, if you do not re-connect the earth you will fail the entire section of exam. If you were to forget this step while working on an installation you will essentially have an un-earthed system that would present a serious risk of electric shock.

Remember: This is a LIVE test and care should be taken at all times during the test procedure.

9.5.3 Earth fault loop impedance of each circuit (Z_s)

Z_s is the measurement of the earth loop impedance of a final circuit including the supply conductor. This test is carried out to ensure that the required disconnection times of each circuit in the installation will be achieved if a fault to earth occurs.

The tests are carried out at the furthest point of each final circuit within the installation. If the measured values of Z_s for each circuit are lower than or equal to the maximum values given in Appendix

B of the *On-Site Guide* (where adjustments for temperature have been made) the required disconnection times would be achieved. See Appendix B of the *On-Site Guide* for more information.

To complete this test you need to follow the steps in order.

ONE: this is a live test so the circuit should be energised at the time of testing.

TWO: prepare the circuit and the test instrument so that each available point can be tested (most test instruments are supplied with special adapters so that they can be plugged into socket-outlets and lampholders).

THREE: in accordance with the manufacturer's instructions for the test instrument, test each accessible point of the circuit and record each reading that you obtain.

FOUR: verify that the highest result obtained is within the required value given in to the IET Wiring Regulations and record it on the Schedule of Test results.

9.5.4 Prospective fault current (I_{pf}) test

Regulation 643.7.3.201 requires that the I_{pf} under both short circuit and earth fault conditions is measured, calculated or determined by another method, such as enquiry.

An PFC test is carried out to find the maximum current that could flow between line and neutral, line and line and between line conductors and earth. An I_{pf} test instrument is part of a twin earth fault loop impedance tester and the result is measured in kA. Regulation 434.1 of the IET Wiring Regulations requires that the prospective fault current should be determined at every relevant point of the installation, i.e. at every point where a protective device is required to operate under fault conditions, including at the origin of the installation. If the I_{pf} is measured at the consumer unit and is found to be satisfactory then it can be assumed that all of the circuits supplied from that consumer unit will also be satisfactory.

This value can be obtained in three ways:

(a) enquiry to the supplier;
(b) calculation; and
(c) measurement.

Measurement of the I_{pf} can be obtained by using a similar method of testing used to determine Z_e, however, the incoming earth should be connected when carrying out this test. The test equipment manufacturer's instructions should be followed so that the correct results can be obtained.

9.5.5 Phase sequence Regulation 643.9

In the case of three-phase installations, you should check that the phase sequencing is correct. This can be done by carrying out a simple test that requires a phase sequence tester. The tester will have three leads generally identified as brown, black and grey or L1, L2 and L3 respectively. In some older installations you will come across conductors identified with red, yellow and blue. the order of sequence is the same The leads should be held on to the corresponding terminals. The rest is automatic. The test instrument will display a reading – similar to that shown in Figure like 9.16.

▼ **Figure 9.16** Typical readings displayed on a phase sequence tester

The image on the left indicates that the phase rotation is in sequence as L1-L2-L3 are running in sequence. The image on the right shows that the phase sequence is not in order indicating that L2 and L3 need to be swapped.

9.5.6 Functional testing of RCDs

The instrument used to carry out this test is an RCD tester. It measures the time it takes for the RCD to interrupt the supply of current flowing through it. The value of measurement is given in either seconds or milliseconds. RCD test instruments generally come supplied with leads that have plug tops for testing socket-outlet circuits and a combination of Bayonet and Edison screw lamp attachments for testing lighting circuits, as well as the standard leads for testing other circuits. Most modern RCD test instruments have an automatic function that carry out the correct sequence of tests, only requiring the person carrying out the test to close the RCD after each operation.

30 mA RCD with integral test button: it has been found in the past that if the RCD has not been operated for some time that this can have an effect on the time it takes for the device to operate. To avoid this and to prevent failure of mechanical operation, the test button must be operated every 6 months.

	Test your knowledge
1.	What can be identified by carrying out an insulation resistance test?
2.	When testing for the continuity of a cpc in a circuit, should the highest or the lowest reading obtained be recorded on the schedule of test results?
3.	What needs to be done with the line and earth conductors when using Method 1 to identify continuity of cpc?
4.	Where can maximum permissible measured values of Z_s be found in the IET's *On-Site Guide*?
5.	When testing for continuity on a lighting circuit, you cannot obtain a low Ω reading. What else might you have to do to test the circuit correctly?
6.	According to GS38 what is the recommended length in mm that the tips of the test probes should be?

Fault Finding   10

After this section, you will have awareness of the following topics:

- ▶ What is a fault?
- ▶ Identifying a fault
- ▶ Fault location
- ▶ Fault rectification

10.1 What is a fault?

Before fault finding can begin, you must know what it is that you are trying to find.

A fault can be a combination of various elements, or a single situation, that prevents an electrical circuit from functioning correctly.

There are generally three types fault that can occur on an electrical circuit.

(a) **Short circuit:** when the length or resistance of a circuit is reduced (shortened), for example, a screw or nail through cable creating continuity between conductors, resulting in an overcurrent.

(b) **Open circuit:** when there is no complete path for current to flow, for example, cable has been removed from the terminal or broken completely.

(c) **Earth fault:** when a fault current flows from a live conductor to earth.

The following examples are not defined as faults in Part 2 of the IET Wiring Regulations, however, awarding bodies include these in exams to assess fault identification and location techniques.

(a) **High resistance:** when the flow of current is restricted, for example, a loose connection, or stranded cable not terminated correctly.

(b) **Cross polarity:** when the line and neutral conductors have been connected the wrong way around.

(c) **Interconnection of conductors:** when conductors have been connected into the wrong terminals, for example, on a heating control unit, the pump and thermostat connected the wrong way around.

Faults and other problems can occur for a number of reasons. Some examples are as follows.

10.1.1 Short circuit

A common fault normally caused by some form of mechanical damage to cable, such as a screw or nail being driven through the conductors or the cable being crushed by furniture or a heavy object. This type of fault will, in most cases, operate the protective device immediately, whether it blows a fuse or trips a circuit-breaker.

10.1.2 Open circuit/high resistance

An open circuit is commonly the result of a break in the circuit as a result of some form of mechanical damage, broken conductor or a component has failed as it has become old and is no longer suitable for continued use or a conductor has simply become loose from its terminal. This will normally result in the equipment not operating at all.

10.1.3 Cross polarity

To put it simply, cross polarity is when the conductors are connected the wrong way around, such as line has been connected into the neutral terminal and the neutral has been connected into the line terminal. This can happen when a job has been rushed or there is insufficient light to correctly identify cable terminals when installing. This is not acceptable and will be identified when the correct test procedures are carried out prior to an installation being energised.

10.1.4 Interconnection of conductors

When installing control systems it can sometimes be difficult to keep track of what conductors you are connecting where. If the cables have not been clearly identified it is possible to get them mixed up. A good example is a heating control system that consists of a number of component parts, such as thermostats, pumps and valves. All the conductors from these parts come back to the main terminal box at the controller. If the conductors have not been identified correctly they may end up in the wrong terminals or if some maintenance work has been carried out they may have been reconnected in the wrong place.

10.1.5 High resistance

High resistance can put additional strain on cables and potentially lead to hazardous situations, such as overheating and fires. This can be a result of damaged electrical equipment, loose connections (high resistive joints), etc.

If a conductor is forced into a terminal that is too small it may damage the terminal, which will potentially result in high resistance.

If a conductor is connected into a terminal that is too large there is a higher risk of the conductor not being sufficiently clamped and it may come loose. All components should be carefully selected so that they serve their intended purpose.

Figure 10.1 shows cable being terminated into connector blocks that are the wrong size. This could result in the cable not being held in the terminal correctly (if you try to use a conductor that is too small) or a connector block being damaged (if you try to use a conductor that is too large).

▼ **Figure 10.1** Incorrect size terminations or connector blocks

10.1.6 Overload

An overload is not a fault – it can be a result of incorrect design, addition of too much current-using equipment, faulty equipment, etc. If the overload current is small it may take some time for the protective device to operate. This can make it a difficult fault to identify and locate. One method of testing for overload is to use a clamp meter. This type of meter can be connected around a conductor to take a reading of the current that is flowing through it. This reading is used in conjunction with the current-carrying capacity of the cable and the rating of the equipment will help to identify if the circuit and any connected items are functioning correctly or not.

10.2 How to diagnose and locate a fault

Before you take any test equipment from your van there is a good chance you can identify the cause of the fault.

Step 1: communication

Communication is key. Speak to the customer to gather as much information as you can. You will need to prompt discussion – remember, not everyone is an electrician, and they may have done something that can affect the electrical installation without realising it.

Hanging a picture, moving some furniture or having had a bathroom fitted recently can all be causes of faults. Alternatively, with changing seasons they may have forgotten about an appliance they only use at certain times of the year, for example, electric heaters may have had a build-up of dust and debris that when switched on has caused damage to the heating elements.

Once you have communicated with the customer and assessed the information available, you can begin further investigation.

Step 2: visual inspection

Once you have communicated with all relevant people about the electrical fault that has been reported, a visual inspection should be carried out.

All electrical appliances and equipment should be removed and disconnected from the installation. This allows the electrician to identify whether or not it is the electrical installation, an appliance or piece of equipment causing the problem.

Most faults are caused by some kind of mechanical impact, such as a nail through cable, damage to electrical equipment that has been moved or replaced, lifed items (items that have reached the end of their functional life).

10.2.1 What to look out for

(a) signs of general wear and tear;
(b) cable not adequately contained or supported;
(c) blackening around terminals and other electrical connection points;
(d) cracked or damaged appliances;
(e) water ingress;
(f) corrosion;
(g) incorrect equipment installed on circuits; and
(h) signs of new construction or installation of electrical and non-electrical equipment, especially wall mounted pictures, shelves etc.

Step 3: identify the type of fault

In order to carry out fault finding safely and to identify the fault, the safe isolation procedure must be carried out (see Section 2). Once this has been done, you can test the circuit. Remember that the only part of the circuit to be tested is the wiring system, so all appliances must be disconnected and lamps removed from fittings.

The following are the types of fault you are most likely to encounter.

10.2.2 Short circuit/earth fault

This fault occurs when two live conductors of the same circuit are touching each other or have been connected by a conductive material – commonly, a nail or screw through a cable.

The first of the tests can be carried out with a low Ohms resistance tester. Check for a short circuit. Hold the probes of the tester across L-N, L-E, and N-E. In a working circuit there should be a high resistance across these conductors.

If the result of the test indicates a low resistance between L-N there is a short circuit. If it indicates a low resistance between L-E or N-E, there is a line to earth or neutral to earth fault .

▼ **Figure 10.2** Nail through cable

As you can see a nail has gone through this cable damaging the outer sheathing as well as the insulation surrounding the copper conductors. As nails are made of a conductive material, electrical current is able to flow from the live conductor through the nail to the neutral conductor, thus creating a path for a short-circuit.

10.2.3 Open circuit

▼ **Figure 10.3** Line conductor damaged

This can be a difficult problem to diagnose, especially on a lighting circuit or circuits with switches, isolators and online starters. One method that can be used is to test the continuity between two conductors as you would when testing the continuity of the cpc on a final circuit. For this test you will need to join the live and cpc together at the consumer unit, locate the end of the circuit and carry out a low Ohms test on all conductors. If the circuit has switches and/or isolators, you will have to carry out this procedure whilst operating all switches. This test needs to be carried out on all conductors, for example, L-N, L-E and N-E. If a low Ohms reading cannot be achieved this suggests that there may be an open circuit fault.

10

10.2.4 High resistance

▼ **Figure 10.4** Poor termination, frayed strands of conductor

This can be the result of a deteriorated termination, poor connection or simply a faulty product or poor installation. This type of fault can potentially cause a fire and may go unnoticed for many years. The best method of avoiding this type of fault is to check and tighten all electrical connections to ensure that they are nice and secure. You can test for this type of fault using a low ohms test instrument; you may find the resistance value higher than expected. To give an indication of what would be a suitable resistance value you can find maximum Z_s values in Table B6 of the IET's *On-Site Guide*.

10.2.5 Cross polarity

▼ **Figure 10.5** Conductors incorrectly terminated

This can be identified fairly easily. When a circuit has been connected incorrectly, in that the live conductor has been connected to the neutral terminal of an accessory or piece of equipment, it can usually be identified visually. If this is not possible, an R1 + R2 type test can be carried out to identify if the conductors are in the correct location. If the L and E are connected together at one end of the circuit you should be able to obtain a low ohms value testing across the same conductors at the other end of the circuit. If a high reading is measured, check N to E, if a low Ohms reading is obtained this indicates that there is cross polarity.

10.2.6 Interconnection of conductors

This type of fault is common in control systems and heating systems. The only way to check for this type of fault is to carry out a continuity test on the individual lengths of conductors while using the manufacturer's circuit diagram as guidance. For example, if the manufacturer's guidance tells you that terminal 12 at the control unit should be connected to terminal 4 at the pump, you will need to test the continuity between the conductors at these two terminals (remove the cables from the terminals). If the reading shows good continuity, this indicates that the conductors have been in the correct location. You will need to continue testing the conductors until you find the one causing the problem.

Step 4: locating the fault

Once the type of fault has been identified it must then be accurately located.

One common method used to identify the location of a fault is a method called sectionalisation or the **half split method**.

This method divides the circuit into smaller sections, which makes it easier to identify the exact location of the fault. The diagrams below show the simple steps that are required to locate a fault on a ring final circuit. This method can also be used on radial circuits.

10.2.7 Finding a fault on a ring final circuit

Figure 10.6 picture shows a ring final circuit. There is a fault on this circuit, somewhere; using sectionalisation the fault can be located with accuracy.

▼ **Figure 10.6** Ring final circuit, no fault

Carry out a risk assessment for the work to be done. Prior to beginning any test, the power supply must be switched off (isolated). This requires the safe isolation procedure to be carried out. When the circuit has been successfully de-energised, the testing procedure can begin:

(a) remove the conductors from terminals at the C/U. Once this has been done you can carry out a test on the conductors and determine what type of fault there is on the circuit.

(b) the first test is to check that the live conductors and earth are electrically separated (as they should be on a normal working circuit). This test will require a continuity test instrument. Carry out the following test procedure:

(i) hold the probes of the tester onto the live and neutral conductors L-N;
(ii) then onto live and earth L-E; and
(iii) then onto neutral and earth N-E.

On a ring final circuit both legs of the ring should be tested.

On a circuit with no faults, the reading on a low Ohms test will provide a high reading (high resistance), for example, >999 MΩ.

▼ **Figure 10.7** Ring final circuit, short circuit between line and neutral

In this example there is a fault between the line and the neutral conductors. When a continuity test is carried out it will indicate a low resistance between these two conductors indicating a short circuit between line and neutral.

The next step is to find out where the fault is. To begin locating the fault, the circuit must be split into two parts (half split method).

As you can see, using this method instantly splits the circuit into two parts, each half the size of the original – only this time, one half of the circuit has no fault.

By carrying out the same test as before you can identify which half that is. The circuit without a fault no longer needs testing and you have essentially reduced the area that needs testing to locate the fault by half.

▼ **Figure 10.8** Ring final circuit split into two parts

If you need to reduce the size of the circuit further, simply repeat the steps. You can continue splitting the circuit into smaller sections, re-testing each part until you have located the fault.

▼ **Figure 10.9** Ring final circuit split again

It may be that the fault is on a length of cable between two socket-outlets. This method will at least reduce the area in which you are fault hunting.

As mentioned earlier in this section, this method can also be used for radial circuits. The principle remains the same: break the circuit down into smaller sections so that the fault can be located.

There are various other ways to find the location of a fault. In most circumstances, it will involve segregating the circuit into smaller parts and testing the individual parts until the fault is located. It is not unusual for an electrician to install new cable to bypass part of a circuit that has a fault as it may be difficult to get to the area of the fault, for example, under a tiled or fixed wooden floor. However, it is the responsibility of the electrician to ensure the safety of an installation before leaving it in the hands of ordinary persons.

Once the fault has been rectified the entire installation that had been affected by the fault must be tested again using the steps described Section 9.

10

Test your knowledge	
1.	If a fault cannot be rectified immediately how must the circuit be left?
2.	When testing earth fault loop impedance, what would indicate that the installation may be satisfactory: a very high value or a very low value?
3.	When fault finding you discover that there is a short circuit on a radial final circuit that supplies 6 BS 1363 socket-outlets. What is the first logical step you would take to begin locating the fault?
4.	What must be done once a fault has been rectified?

Common Calculations 11

After this section, you will have awareness of the following topics:

- Simple transposition
- Triangle method
- Resistors in series
- Resistors in parallel

11.1 Simple transposition

11.1.1 What is transposition?

When you need to calculate a value using an equation, you may need to transpose the formula in order to find the unknown. Calculations have rules; if you do something with one value it will affect another, for example, if you increase the resistance of a circuit you will reduce the amount of current (I amps) that can flow through that circuit. By moving the values around in the calculation the answer can be found. The following rules must be applied so that transposition can be used correctly:

(a) the unknown should always be on its own, on the top line and on the left-hand side of the equation;

(b) to get the unknown on its own some of the values must be moved; and

(c) when moving a value across the equals sign it must move either from above to below the line or from below to above the line.

For example:

3 × 6 = 9x

The first step is to move the unknown to the left-hand side of the equation:

9x = 3 × 6

Then draw an imaginary line underneath the right hand side of the equation. Remember the unknown is always on the top line:

9x = 3 × 6

To transpose this you have to leave the unknown value on its own, so the 9 will be the logical value to move:

$$9x = \frac{3 \times 6}{9}$$

When the 9 is moved to the opposite side of the equals sign it must go below the line, changing the equation to this:

$$x = \frac{3 \times 6}{9}$$

Once you have done the maths you can find the answer:

$$x = \frac{18}{9}$$

x = 2

To check the answer, you can re-write the equation to look like this:

3 × 6 = 9 × 2

3 × 6 = 18 and 9 × 2 = 18

Remember that one side of an equation will always be equal to the other side!

11.2 Triangle method: voltage, amps, resistance and watts

The following triangles will help when calculating simple values for voltage, amps, resistance and watts.

They are the first of many formulae you will learn as you progress in your electrotechnical career.

| V for voltage in volts (V) | I for amperes in amps (A) |
| R for resistance in Ohms (Ω) | P for power in watts (W) |

You can use the triangles below by covering the value you want to find. For example, if you already know the value of V and R and you want to find I you simply cover I on the triangle. This will indicate that you need to divide V by R, which will give you the value for I. The same principle applies when trying to find any other values.

 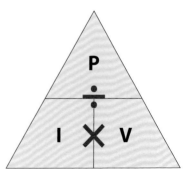

Ohm's Law (VIR)	Power Law (PIV)
In this triangle you can see that: R = V ÷ I I = V ÷ R V = I × R	In this triangle you can see that: V = P ÷ I I = P ÷ V P = I × V

11.3 Resistors

Note: It's important to use the correct terminology when citing voltage and current, i.e. current is always 'through' a component and voltage is always 'across' a component.

11.3.1 Resistors in series

In a series circuit, the current remains constant through each resistor. If, for example, there is 2 A through R_1, there will be 2 A through R_2 and 2 A through R_3.

When adding up the total resistance of a series circuit you simply add all the values together.

To calculate the total resistance of the following circuit, add all the resistor values together:

$$R_t = R_1 + R_2 + R_3$$
$$R_t = 3 + 3 + 3$$
$$R_t = 9$$

So the total resistance (R_t – t for total) is 9 Ω.

11.3.2 Resistors in parallel

In a parallel circuit, the voltage remains constant across each resistor. If, for example, there is a 20 V supply, there will be 20 V across R_1 and 20 V across R_2.

When resistors are connected in parallel an extra path is being created for the current to flow. So, instead of increasing the overall resistance, as happens in a series circuit, the overall resistance is actually reduced.

Imagine the resistors are narrow alleyways and the current is a group of people trying to get through. If there is only one alleyway it will take twice as long for a group of people to walk through than it would if there were two narrow alleyways.

To calculate the total resistance of a resistor network connected in parallel you need to establish the current through each resistor. The voltage remains constant across each resistor in a parallel circuit.

Remember Ohm's Law: $I = \dfrac{V}{R}$

Which looks like: $I = \dfrac{20}{5}$ ← 20 V R_1 5 Ω R_2 5 Ω

$I = 4$

So the current flow through each resistor is 4 A.

To calculate the total current for the circuit the current flow through each resistor must be added together. In the example above it would be:

4 + 4 = 8 A

So a total of 8 A is flowing through the circuit. Now that we have the total current for the circuit we can work out the total resistance of the circuit using Ohm's law:

$\dfrac{20}{8} = 2.5$ Ω

You can see that the total resistance is reduced when an additional resistor is connected in parallel in the circuit.

What if we don't know the value of V?

There is another way of working out the total resistance of resistors in parallel. If only the resistance values are known you can use the following equation.

$$\dfrac{1}{R_t} = \dfrac{1}{R_1} + \dfrac{1}{R_2}$$

$$\frac{1}{R_t} = \frac{1}{5} + \frac{1}{5}$$

$$\frac{1}{R_t} = \frac{1+1}{5}$$

$$\frac{1}{R_t} = \frac{2}{5} = 0.4$$

Remember the value you are trying to find is not a fraction so you therefore need to carry out one more calculation: If

$$\frac{2}{5}$$

is equal to

$$\frac{1}{R_t}$$

then

$$\frac{R_t}{1}$$

is equal to

$$\frac{5}{2}$$

so this essentially means that

$$R_t = \frac{5}{2}$$

$R_t = 2.5 \; \Omega$

When using a calculator you will need to use the X^{-1} button and the S↔D button.

The sequence will look like this:

[5] → [x^{-1}] → [+] → [5] → [x^{-1}] → [=] → [2/5] → [SD] → [0.4] → [x^{-1}] → [=] → [2.5]

In this example, the answer is:

2.5 Ω

This answer should always be a lower resistance than the smallest resistor on the circuit.

Finding a percentage (%) the easy way

Let's say you want to determine what 4 % V drop is on a circuit supplied at 230 V. The easiest way to find out looks like this:

$$VD = \frac{V \times \%}{100}$$

$$\frac{230 \times 4}{100} = 9.2$$

When using the % function on a calculator you should enter the values like this:

11

Test your knowledge

1.	Try transposing these:
	$4 \times 6 = 2x$ $9 \times 3 = 3x$ $4 \times 2 \times 3 = 2 \times 4x$
2.	What is power measured in?
3.	If you have the values of R and I what must you do to find V?
4.	If a 230 V electrical appliance is rated at 460 W, what will the current flow be in amps when the electrical supply is 230 V?
5.	Calculate the total resistance (R_t) of these circuits:

Circuit 1: 6 Ω — 3 Ω — 0.5 Ω $R_t =$

Circuit 2: 12 Ω — 30 Ω — 9 Ω $R_t =$

Circuit 3: 6 Ω — 2 Ω — 7 Ω $R_t =$

6.	Calculate R_t in these circuits:

20 V, parallel: 7 Ω, 3 Ω $R_t =$

25 V, parallel: 9 Ω, 2 Ω, 1 Ω $R_t =$

Diversity 12

After this section, you will have awareness of the following topics:

- **What diversity is and why we need it**
- **Calculating diversity**
- **Applying diversity to various installations**

12.1 What is diversity?

Diversity is the term used to describe the calculation that determines the likely maximum demand at any one point on an electrical circuit or installation (not to be confused with the maximum connected load). This enables electricians to design an electrical installation with the assumption that every piece of electrical equipment will not be switched on and running at full capacity all the time. Cables can withstand high currents for short periods of time and in some cases, for example, when designing a circuit for a motor or for equipment that uses a large amount of power for a short period of time, a smaller cross-sectional area can be used.

If every installation was designed to carry the maximum connected load the cost to customers would be extremely high and the wiring systems over-engineered. In some ways, this is similar to the way in which a car is designed and manufactured as the vehicle is not intended to be driven at maximum speed everywhere it goes. Similarly, it would be unrealistic to assume that all electrical equipment will be switched on at full power all the time.

12.2 Calculating diversity

Because diversity is worked out differently depending on the characteristics of the circuit, the easiest way to explain how to calculate diversity is through an example. When designing an installation to supply a cooker hob and oven, the following method can be used.

The oven is rated at 4,000 W and you have four rings on the hob:

- 800 W
- 1,000 W × 2
- 2,000 W

Looking at these values the assumed design current would be:

800 + (1,000 × 2) + 2,000 + 4,000

Total connected load = 8,800 W (8.8 kW)

We know that watts is the measurement of power so, by using the power triangle from Section 11, the calculation to determine the design current

$$I_b = \frac{V}{P}$$

would look like this:

$$I_b = \frac{8800}{230}$$

so:

$I_b = 38.26$ A

However, this is the total connected load of the hob and oven when every element is turned. The likeliness of this happening for a long period of time is extremely unlikely and if the electrician were to install a cable to meet these requirements, the cable would be oversized and only ever reach a fraction of its current carrying capacity, as well as costing the customer a lot more than necessary. If a protective device is chosen based on these requirements it may not be suitable for its intended purpose.

Using Tables A1 and A2 from the IET's *On-Site Guide*, it is possible to determine roughly how much actual current the installation will be drawing, known as diversified load.

▼ **Figure 12.1** Extract from Table A1 *On-Site Guide*

Point of utilisation or current-using equipment	Current demand to be assumed
Socket-outlets other than 2 A socket-outlets and other than 13 A socket-outlets See note 1	Rated current
2 A socket-outlets	At least 0.5 A
Lighting outlet See note 2	Current equivalent to the connected load, with a minimum of 100 W per lampholder
Electric clock, shaver supply unit (complying with BS EN 61558-2-5), shaver socket-outlet (complying with BS 4573), bell transformer, and current-using equipment of a rating not greater than 5 VA	May be neglected for the purpose of this assessment
Household cooking appliance	The first 10 A of the rated current plus 30 % of the remainder of the rated current plus 5 A if a socket-outlet is incorporated in the control unit
All other stationary equipment	British Standard rated current, or normal current

Using the table we can begin to calculate the maximum demand (connected load including an allowance made for diversity).

For the calculation of the hob and oven: the design current is 38.26 A (from our earlier calculation), with the first 10 A taken at 100 % of its value. This means that you take 10 A from the value and put it to the side ready to add to the remaining value once the following calculations have been carried out.

So:

38.26 − 10 = 28.26 A

The remaining 28.26 A is taken at 30 %. The easiest way to work out what 30 % of 28.26 is:

(28.26 ÷ 100) × 30 (or, if you have a calculator (28.26 × 30 %)) = 8.47

Once this value has been calculated it can be added to the 10 A that was put on the side at the beginning: 8.47 + 10 = 18.47 A.

▼ Figure 12.2 Process sequence for calculating diversity for a cooker circuit

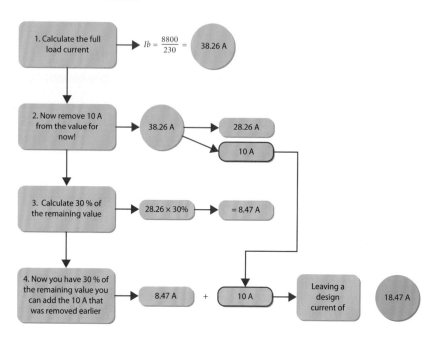

What next?

The value remaining after the calculation has been completed is used to select the rating of the protective device. In this case, the value is 18.47 A, so a protective device of this rating or the next available higher rating must be selected, for example, a 20 A circuit-breaker (see the relevant Standard, BS EN 60898).

12.3 Diversity for ring final circuits

Each socket-outlet is designed for a plug of up to 13 A, for a double socket that is potentially 26 A. In the circuit shown below the total current that may be drawn from this circuit is 130 A.

Figure 12.3 Ring final circuit

It would be unrealistic to install cable and circuit-breakers with a rating of 130 A for a ring final circuit as it is extremely unlikely that it will ever be used to this extent, so diversity can be taken into account. Table H2.1 from the IET's *On-Site Guide* provides guidance on the maximum floor area that can be served with BS 1363 socket-outlets. Diversity has already been taken into account in this table and does not need to be applied again.

12

▼ **Figure 12.4** Table H2.1 from the IET's *On-Site Guide*

Type of Circuit		Overcurrent protective device rating (A)	Minimum live conductor cross-sectional area* (mm²)		Maximum floor area served (m²)
			Copper conductor thermoplastic or thermosetting insulated cables	Copper conductor mineral insulated cables	
1	2	3	4	5	6
A1	Ring	30 or 32	2.5	1.5	100
A2	Radial	30 or 32	4	2.5	75
A3	Radial	20	2.5	1.5	50

Kitchen appliance demands

The image below shows how diversity can be considered for circuits that are designed to supply various appliances that have a varying load profile depending on the stage of its cycle. For example, the washing cycle of a washing machine begins with it turning to take on water and wet the garments within. It then heats the water, has various washing and rinsing cycles, ending with a long spin cycle to remove the water. The appliance, therefore, will not run at maximum power for the entire duration of the wash.

You will notice on the image that the total load increases to above 30 A on two occasions for a short period. Most protective devices will allow this small overload current for a limited time and it is considered an acceptable practice when designing an installation. In the circuit below a 32 A circuit-breaker (BS EN 60898) can therefore be considered for a protective device as the load is only expected to be higher than 32 A for a very short time.

▼ **Figure 12.5** Diversity used in different circuits

When can diversity not be applied?

When designing an installation for equipment, such as a single water heater (thermostatically controlled) or an electric shower, the design current will be constant and diversity should therefore not be applied. This type of circuit will be either 100 % on or 100 % off. This means that, when a water heater is switched on, it will use the same amount of power until it is switched off again. The demand will not vary as it does with cooker circuits and washing machines etc.

The important thing to remember about diversity is that it is specific to every installation and the electrician/designer must consider what the likely demand is going to be on the circuit when selecting the appropriate cable and protective devices.

12

	Test your knowledge
1.	You have an appliance that, when operating normally, will draw a current of 15.5 A. What rating should the overcurrent protective device be?

2.	If diversity has not been taken into consideration, what are the implications on the:
	   (a) installer?
	   (b) customer?

3.	From Table A2 from the IET's *On-Site Guide*, what is the recommended value, with diversity taken into consideration, for cooking appliances in small shops, stores, offices and business premises?
4.	Name two electrical appliances that cannot have diversity applied to them?
5.	On a lighting circuit, how many watts per light fitting should be taken into consideration when working out diversity?

6.	You have a 230 V hob and oven, in a domestic property, with:
	   ▶ two rings on the hob with a rating of 2,000 W;
	   ▶ two rings on the hob with a rating of 750 W; and
	   ▶ the oven with a rating of 3,000 W with an additional socket-outlet.
	What size circuit-breaker should be installed? Show all working.

Special Locations

A

As mentioned at the beginning of this Guide, the IET Wiring Regulations have been around since 1882. It goes without saying that a lot has changed since then. Electricity has taken over almost every possible location there is, such as solar photovoltaic systems on roof tops; swimming pools; electric vehicle charging points on the street or in domestic dwellings, etc.

Even 50 years ago it was impossible to imagine the extent that electricity would be used today. Part 7 of the IET Wiring Regulations takes into consideration some of the special installations or locations where electrical systems will be installed. It is not the intention to create completely new regulations to suit various different locations but rather to identify which existing regulations need to be considered and, if necessary, modified when installing an electrical system in a special location. For one reason or another, all of the special installations or locations mentioned in Part 7 of the IET Wiring Regulations need additional consideration when electrical equipment and wiring systems are being installed. It needs to be borne in mind that Part 7 should be used in conjunction with the other parts of the IET Wiring Regulations and not as the sole requirements for that installation.

In this Annex, we only briefly introduce you to the areas that qualify as special locations for installations. The IET's Guidance Note 7 Special Locations provides further information and guidance on all the special locations and installations as set out in the IET Wiring Regulations. It is recommended that you consult this Guidance Note and the IET Wiring Regulations should you find yourself working on any of the areas listed in the below table.

There are currently 19 special installations or locations covered in Part 7.

Location	Section	Description
Locations containing a bath or shower	701	What needs to be borne in mind is that we conduct electricity more easily when we come into contact with a live part when we are wet. The following factors should be considered when installing electrical systems in a location containing a bath or shower: (a) it is likely that persons will not be wearing any clothing or footwear – this increases the contact area they have with surfaces at earth potential. (b) the presence of water, which can reduce the contact resistance between parts that may become live (under fault conditions) and the human body. (c) when people are immersed in water the total body resistance to current is reduced – more information on the resistance of the body can be found in IEC/TS 60479. Figures 701.1 of the IET Wiring Regulations and 8.1 (i) and (ii) of the *On-Site Guide* show the zone dimensions in a location containing a bath or shower.
Swimming pools and other basins	702	As mentioned in 701 above, the resistance between the body and earth is reduced when a person is unclothed or partially unclothed, especially when not wearing some form of footwear and when the body is likely to be wet. 702 applies to the basins of swimming pools, fountains and paddling pools. It also covers the surrounding area of these basins that have allocated zones.

Location	Section	Description
Rooms and cabins containing sauna heaters	703	If there are showers or similar facilities installed, the requirements of 701 also apply. Temperature and extreme humidity are factors that really need to be considered in locations containing sauna heaters. 703 provides the installation requirements and the zone dimensions.

Location	Section	Description
Construction and demolition site installations	704	Construction sites are potentially dangerous in many ways. Four factors contribute to the high risk of electric shock on a construction site: **(a)** the possibility of damage to cables and equipment. **(b)** the widespread use of hand tools with trailing leads (this problem is mitigated by the increasing use of battery operated tools). **(c)** the accessibility of many extraneous-conductive-parts, which cannot practically be bonded. **(d)** the works are generally open to the elements. Measures to reduce the risks include: **(a)** the preference, as stated in the IET Wiring Regulations, for the reduced low voltage system (110 V AC which, at single phase, is 55 V to earthed midpoint of the transformer) to supply portable handlamps for general use and portable hand tools and local lighting up to 2 kW, while SELV is strongly preferred for portable handlamps in confined or damp locations. **(b)** the avoidance of using a PME (TN-C-S) earthing facility for the means of earthing for a construction site installation unless all extraneous-conductive-parts are reliably connected to the main earthing terminal. **(c)** protective measures of obstacles, placing out of reach, non-conducting location and prohibiting or not permitting earth free local equipotential bonding. **(d)** requiring that equipment for external use should be at least IP44. **(e)** additional testing of RCDs.

Location	Section	Description
Agricultural and horticultural premises	705	705 provides the requirements for agricultural and horticultural premises, such as farmyards, cattle yards, slaughterhouses, greenhouses, etc. In circuits, whatever the type of earthing system, the following protective devices shall be provided: **(a)** circuits supplying socket-outlets up to 32 A: a 30 mA RCD. **(b)** circuits supplying socket-outlets over 32 A: a 100 mA RCD. **(c)** all other circuits: a 300 mA RCD.

Location	Section	Description

Conducting locations with restricted movement	706	The particular requirements of 706 apply to fixed equipment in conducting locations where movement of persons is restricted by the location, and supplies to mobile equipment in these locations. Such locations are where movement is heavily restricted and the area is metallic and electrically conductive. Examples of this type of location are working at high level on a steel walkway or gantry or within a stainless steel storage vessel.
		
		There are particular requirements for protection against electric shock. In conducting locations with restricted movement the following protective measures apply to circuits supplying the following current-using equipment:
		
		(a) supply to a hand-held tool or an item of mobile equipment:
		    (i) electrical separation; or
		    (ii) SELV.
		(b) supply to handlamps: SELV.
		(c) supply to fixed equipment:
		    (i) automatic disconnection with supplementary equipotential bonding; or
		    (ii) by use of Class II equipment and additional protection by the use of RCDs at 30 mA; or
		    (iii) electrical separation; or
		    (iv) SELV; or
		    (v) PELV.

Location	Section	Description
Marinas and similar locations	709	The environment of a marina or yachting harbour is harsh for electrical equipment. The water, salt, high winds and movement of structures accelerate deterioration of the installation. The presence of salt water, dissimilar metals and a potential for leakage currents increases the rate of corrosion. There are also increased electric shock risks associated with a wet environment, by reduction in body resistance and contact with earth potential. Particular requirements to reduce the risks include: **(a)** prohibition of a TN-C-S system for the supply to a boat. **(b)** additional protection by 30 mA RCDs. **(c)** socket-outlets to be installed at not less than 1 m above the highest water level (may be reduced if additional measures are taken).

Location	Section	Description
Medical locations	710	710 provides requirements for hospitals, private clinics, medical and dental practices, healthcare centres, dedicated medical rooms in the work place and can be used in veterinary clinics.

Risks in this location may arise from the following factors:

(a) the natural protection of the human body against electric shock is considerably reduced when certain clinical procedures are being performed on it.
(b) patients under treatment may have their skin resistance broken or their defensive capacity either reduced by medication or nullified while anaesthetised.
(c) in addition to the risk from electric shock, some equipment, such as life-support equipment or surgical equipment, performs such vital functions that loss of the electrical supply would pose an unacceptable risk to patients. Medical locations, where such vital equipment is used, require secure electrical supplies.

To protect patients from possible electrical hazards, 710 requires additional protective measures to be applied in medical locations.

These include:
(a) particular requirements for protection against electric shock;
(b) medical IT systems (see Guidance Note 7);
(c) requirements for supplementary equipotential bonding;
(d) additional requirements for the selection and erection of electrical equipment including switchgear and controlgear;
(e) safety services including the sources, such as generators; and
(f) detailed requirements for safety lighting.

A

Location	Section	Description
Exhibitions, shows and stands	711	711 applies to temporary electrical installations in exhibitions, shows and stands.

It does not apply to:

(a) the building in which the event is occurring;
(b) electrical systems used in the structures, sets, etc., of production and similar activities of the entertainment industry; and
(c) the exhibits themselves.

The risks associated with exhibitions, shows and stands are those of electric shock and fire. These arise from:

(a) the temporary nature of the installation;
(b) lack of permanent structures;
(c) severe mechanical stresses; and
(d) access to the general public.

Because of these increased risks, additional measures are required:

(a) a cable intended to supply temporary structures shall be protected at its origin by an RCD with maximum residual operating current 300 mA.
(b) bonding of all metallic structural parts that are accessible within the stand etc.
(c) additional protection is required for all final circuits (other than for emergency lighting) and sockets-outlets up to 32 A by RCD, in accordance with Regulation 415.1.1.
(d) TN-C-S shall not be used (there are exceptions).
(e) every temporary structure, such as a vehicle, stand or unit, intended to be occupied by one specific user and each distribution circuit supplying outdoor installations shall be provided with its own readily accessible and properly identifiable means of isolation.

Location	Section	Description
Solar photovoltaic (PV) power supply systems	712	712 applies to the electrical installations of PV power supply systems. Some key points: **(a)** PV modules generate electricity at DC, it is connected to an inverter that converts DC to AC to allow the connection to the AC electrical supply of the building. The DC side of the installation, i.e. the PV modules, is considered to be energised, even when the system is disconnected from the AC side. Remember that when it's daylight, the modules are generating. **(b)** to allow for maintenance, the means of isolating the AC and DC sides of the PV convertor shall be provided. **(c)** there are also many other additional requirements (see 712 in the IET Wiring Regulations).
Outdoor lighting installations	714	Outdoor lighting installations include those for roads, parks, car parks, gardens, places open to the public, sporting areas, illumination of monuments and floodlighting. In addition, other lighting arrangements in places such as telephone kiosks, bus shelters, advertising panels and town plans are covered and all require RCD protection. Road signs are also included but not road traffic signal systems. 714 does not include temporary festoon lighting, chiefly because it will be erected on an adhoc basis and will not form part of the fixed installation. Furthermore, luminaires fixed to the outside of a building and supplied directly from the internal wiring of that building are not included as these are considered part of the building's electrical installation.

Location	Section	Description
Extra-low voltage lighting installation	715	The particular requirements of these regulations apply to extra-low voltage lighting installations supplied from sources with a maximum rated voltage of 50 V AC rms or 120 V DC. The regulations include requirements for: (a) protection against electric shock (SELV); (b) protection against the risk of fire due to short circuit; (c) types of wiring systems, including special requirements where bare conductors are used; (d) the types of transformers and converters; and (e) requirements for suspended systems.

A

Location	Section	Description
Mobile or transportable units	717	The particular requirements of 717 apply to mobile or transportable units. These may be self propelled, towed or transportable containers or cabins. Examples of the units include technical and facilities vehicles for the entertainment industry, medical services, advertising, fire fighting, workshops, offices, transportable catering units, etc.

The risks associated with mobile and transportable units include:

(a) risk of loss of connection to earth due to use of temporary supply cables,

(b) risks arising from the connection to different national and local electricity distribution networks;

(c) open-circuit faults of the PEN conductor of PME (TN-C-S) supplies, raising the potential of all metalwork (including that of the unit) to dangerous levels;

(d) risk of shock arising from high functional currents flowing in protective conductors; and

(e) vibration while the vehicle or trailer is in motion, or while a transportable unit is being moved – causing faults within the unit installation.

Some of the requirements to reduce these risks include:

(a) additional protection by an RCD.

(b) accessible conductive parts of the unit to be connected through the main equipotential bonding to the main earth terminal within the unit.

(c) identification adjacent to the supply inlet:
  (i) the type of supply that may be connected.
  (ii) the voltage rating of the unit.
  (iii) the number of phases and their configuration.
  (iv) the onboard earthing arrangement.
  (v) the maximum power required by the unit.

Location	Section	Description
Electrical installations in caravan/ camping parks, caravans and motor caravans	708 and 721	The particular requirements of 708 apply to the electrical installations in caravan/camping parks and similar locations providing connection points for supplying leisure accommodation vehicles (including caravans) and tents. The particular requirements of 721 apply to the electrical installations of caravans and motor caravans at nominal voltages not exceeding 230/440 V AC or 48 V DC. Note there are some exclusions. The risks specifically associated with installations in caravan parks, caravans and motor caravans arise from: **(a)** open circuit faults of the PEN conductor of PME (TN-C-S) supplies raising the potential to true Earth of all metalwork to dangerous levels. **(b)** incorrect polarity at the pitch supply point. **(c)** possible loss of earthing due to long supply cable runs, connecting devices exposed to weather and flexible cable connections liable to mechanical damage. **(d)** vibration while the vehicle is moving, causing faults within the caravan installation. Particular requirements to reduce the above risks include: **(a)** prohibition of the connection of exposed- and extraneous-conductive-parts of a caravan or motor caravan to a PME (TN- C-S) terminal. Where the supply to the site is PME (TN-C-S), the earthing arrangement at the pitch supply point shall form part of a TT system. **(b)** additional protection by 30 mA RCDs in both the vehicle and the pitch supply point. **(c)** double-pole isolating switch and circuit-breakers protecting final circuit in the tent, caravan or motor caravan. **(d)** internal wiring of the tent, caravan or motor caravan by flexible or stranded cables of cross-sectional area 1.5 mm^2 or greater; additional cable supports; segregation of low voltage and extra-low voltage circuits.

A

Location	Section	Description
Electric vehicle charging installations	722	722 specifies the requirements for charging supplies to electric vehicles. It includes requirements for the type and current rating of socket-outlets, RCD protection, measures of protection against electric shock, earthing requirements, IP rating of equipment, impact protection against mechanical damage, isolation and switching and fixing arrangements, etc. An important point is that Regulation 722.411.4.1 does not allow PME (TN-C-S) as a means of earthing for an electric vehicle charging point where the charging point or the vehicle is located outdoors, except where dangerous touch voltages are not possible or will be disconnected before a dangerous voltage occurs or a supplementary earth electrode is installed. There is an exception for a dwelling if none of the preventative measures is reasonably practicable. It is worth noting that in this regulation the meaning of 'dwelling' is restricted to a self-contained unit designed to accommodate a single household. Another key point is that socket-outlets must be fit for purpose. They must be suitable for the load, and for the external influences such as protection against mechanical damage and ingress of water. 722 requires a degree of protection of at least IP44 (protected against solid objects over 1 mm and protected against water splashed from all directions) where the equipment is installed outdoors. The use of 13 A extension leads, for example, would not meet the requirements of 722, however, the use of tethered vehicle connectors is allowed.

Location	Section	Description
Operating and maintenance gangways	729	729 applies to basic protection and other aspects in restricted access areas, for example, switchrooms with switchgear and controlgear assemblies with a need for operating or maintenance gangways for authorised persons. 729 contains requirements for: **(a)** restricted access areas where basic protection is provided by barriers or enclosures. **(b)** restricted access areas where the protective measure of obstacles is applied. 729 gives examples of positioning doors in long closed restricted access areas.
Onshore units of electrical shore connections for inland navigation vessels	730	Section 730 applies to onshore installations that are dedicated to the supply of inland navigation vessels for the commercial and administrative purposes, berthed in ports and berths. Most, if not all of the measures used to reduce the risks in marinas apply equally for electrical shore connections for inland navigation vessels, One of the major differences between supplies to vessels in a typical marina and electrical shore connections for inland navigation vessels is the size of the supply needed. For example, vessels used on inland waterways in Europe can be up to 10, 000 tonnes, which are considerably larger than the average size of vessel used in a marina, which are generally small recreational craft (up to 24 m long).

Location	Section	Description
Temporary electrical installations for structures, amusement devices and booths at fairgrounds, amusement parks and circuses	740	740 specifies the minimum electrical installation requirements to facilitate the safe design, installation and operation of temporary erected mobile or transportable electrical machines and structures that incorporate electrical equipment. The machines and structures are intended to be installed repeatedly, without loss of safety, in a temporary capacity, at fairgrounds, amusement parks, circuses or similar places. The permanent electrical installation is excluded from the scope. 740 does not apply to the internal electrical wiring of the machines.

Location	Section	Description
Floor and ceiling heating systems	753	753 applies to the installation of electric floor and ceiling heating systems. It does not apply to wall heating or outdoor heating systems.
		The risks associated with ceiling heating systems are generally that of penetration of the heating element by nails, drawing pins, etc. pushed through the ceiling surface. For this reason 753 requires that RCDs with a maximum rated residual operating current of 30 mA shall be used for automatic disconnection of supply.
		Similarly, there are concerns that under-floor heating installations can be damaged by carpet gripper nails, etc. and for similar reasons protection by a 30 mA RCD is required.
		To protect the building structure and provide precautions against fire, there are requirements to avoid overheating of the floor or ceiling heating system.
		In floor areas where contact with skin or footwear is possible, the surface temperature of the floor must be limited (for example, 35 °C).
		Regulation 753.411.3.2 requires that heating units manufactured without exposed-conductive-parts to be provided on site with a grid with spacing of not more than 30 mm, or other suitable conductive covering above the floor heating or below the ceiling heating and connected to the protective conductor of the installation.

A

Tables of Symbols   B

▼ **Table B.1** General symbols

V	volts
A	ampere
I	current
Ω	Ohm
Hz	hertz
W	watts
kW	kilowatts
C	capacitance
F	farads
uF	microfarad
p.u.	per unit
ph	phase
p.f.	power factor
L	line
N	neutral
h	hour
min	minutes
s	second
⚌	direct current (DC)
∿	alternating current (AC)

B

Symbol	Description
2 ∿	two-phase alternating current
2N ∿	two-phase alternating current with neutral
3 ∿	three-phase alternating current
3N ∿	three-phase alternating current with neutral
IPXX	IP number
⚡	fault (indication of assumed fault location)
▢	Class II appliance. Equipment with this mark indicates that additional measures have been taken to prevent the risk of the user receiving an electric shock. Commonly, an additional layer of insulated material protecting live parts (see BS 2754).
◇III◇	Class III appliance (equipment in which protection against electric shock relies on supply at SELV and in which voltages higher than those of SELV are not generated) (see BS 2754).
(safety isolating transformer symbol)	Safety isolating transformer. Class III equipment must be supplied from a safety isolating transformer to BS EN 61558-2-6. The safety isolating transformer will have this identifying mark upon it.
(isolating transformer symbol)	Isolating transformer
⏚	Protective earth, general symbol (preferred to ⏟)

B

▼ **Table B.2** International system of units: base units

In the International System of units (known as SI), there are seven base units, as shown below.

Quantity	Unit	Name of base unit	Unit symbol
Length	L	metre	m
Mass	m	kilogram	kg
Time	t	second	s
Electric current	I	ampere	A
Thermodynamic temperature	degree	kelvin	K
Amount of substance	-	mole	mol
Luminous intensity	Iv	candela	cd

▼ **Table B.3** Multiples and sub-multiples of quantities*

10^{18}	exa	E				10^{-3}		milli	m
10^{15}	peta	P	10^{2}	hecto	h	10^{-6}		micro	μ
10^{12}	tera	T	10^{1}	deca	da	10^{-9}		nano	n
10^{9}	giga	G	10^{-1}	deci	d	10^{-12}		pico	p
10^{6}	mega	M	10^{-2}	centi	c	10^{-15}		femto	f
10^{3}	kilo	k				10^{-18}		atto	a

* Powers in steps of 3 are preferred but some others have common usage (e.g. centimetre, cm; decibel, dB).

▼ **Table B.4** SI derived units, extracted from J1.1

The units of all physical quantities are derived from the seven base units and supplementary SI units and certain of them have been named. These, together with some common compound units, are given here:

Quantity		Unit name	Unit symbol	SI units
Force	F	newton	N	kg m/s^2
Energy	E	joule	J	N m
Power	P	watt	W	J/s
Pressure, stress	p	pascal	Pa	N/m^2
Electric potential	V	volt	V	J/C, W/A
Electric charge, electric flux	Q	coulomb	C	A s
Magnetic flux	ΦB	weber	Wb	V s
Magnetic flux density	B	tesla	T	Wb/m^2
Resistance	R	ohm	Ω	V/A
Conductance	G	siemens	S	A/V
Capacitance	C	farad	F	C/V
Inductance	L	henry	H	Wb/A
Celsius temperature	c	degree Celsius	°C	K
Frequency	f	hertz	Hz	s^{-1}
Luminous flux	Φ_V	lumen	lm	cd sr
Illuminance	E_V	lux	lx	lm/m^2
Mass density	m	kilogram per cubic metre	kg/m^3	
Torque	T	newton metre	N m	
Electric field strength	E	volt per meter	V/m	
Magnetic field strength	H	ampere per metre	A/m	
Thermal conductivity	K	watt per metre kelvin	W m^{-1}K^{-1}	
Luminance	sb	candela per square metre	cd/m^2	

▼ **Table B.5** Symbols for use in schematic wiring diagrams

Symbol	Description
(transformer symbol)	Transformer, general symbol
(wiring with joint symbol)	General wiring, with joint
Phase(s) ⫲ N ⟋ Protective conductor ⟋ Combined protective and neutral conductor (PENT) ⟋	identification of specific conductors
(coil symbol)	Operating device, general symbol (coil)
(make contact symbol)	Make contact, normally open
(break contact symbol)	Break contact, normally closed
(manual switch symbol)	Manually operated switch, general symbol
△	Three-phase winding, delta
(diode symbol)	Semiconductor diode, general symbol
Y	Three-phase winding, star

◰	Changer, general symbol Convertor, general symbol **Notes:** (a) If the direction of change is not obvious, it may be indicated by an arrowhead on the outline of the symbol (b) A symbol or legend indicating the input or output quantity, waveform, etc., may be inserted in each half of the general symbol to show the nature of the change.
	Rectifier
	Inverter
	Battery of primary or secondary cells
	Fuse link, rated current in amperes, general symbol
	Capacitor, general symbol
	Inductor coil, winding or choke
	Inductor coil, winding or choke with magnetic core

▼ **Table B.6** Making and breaking current

	Switch
	Switch-fuse
	Fuse-switch

B

▼ **Table B.7** Isolating, extracted from J2.2

Symbol	Description
	Isolator (disconnector), general symbol
	Disconnector-fuse (fuse combination unit)
	Fuse-disconnector
	Circuit-breaker suitable for isolation

▼ **Table B.8** Making, breaking and isolating, extracted from J2.3

Symbol	Description
	Switch-disconnector
	Switch-disconnector-fuse (fuse combination unit)
	Fuse-switch-disconnector

▼ **Table B.9** Meters

Symbol	Description
(V)	Voltmeter
(A)	Ammeter
*	Integrating instrument or energy meter * Function Wh = watt-hour VArh = volt ampere reactive hour

▼ **Table B.10** Location symbols for installations

Symbol	Description
⊛	Machine, general symbol * Function M = Motor G = Generator
	Motor starter, general symbol
	Star-delta starter
	Socket-outlet, general symbol
	Switched socket-outlet
	Switch, general symbol
	2-way switch, single-pole
	Intermediate switch
	Pull switch, single-pole

	Lighting outlet position, general symbol
	Fluorescent luminaire, general symbol
	Emergency lighting luminaire (or special circuit)
	Self-contained emergency lighting luminaire
	Push button with indicator lamp
	Lamp or signal lamp, general symbol
	Clock, general symbol
	Acoustic signalling device, general symbol (e.g. bell)
	Buzzer

B

Symbol	Description
⌂	Telephone handset, general symbol
◁	Microphone
◁⏐	Loudspeaker
Y	Antenna, general symbol
[G]	Static generator

Answers

Section 1 The IET Wiring Regulations

1. They were first issued in 1882.
2. The international committee responsible for electrotechnical regulations is the International Electrotechnical Committee (IEC).
3. 'Second fix' is one of the final stages of an electrical installation carried out near the end of construction.
4. Appendix 4 contains information about the various types of installation methods.
5. The UK committee responsible for the IET Wiring Regulations is the Joint Power Electrotechnical committee (JPEL/64).
6. The term used is 'chasing'.
7. Electrical Safety Quality and Continuity Regulations (ESQCR) 2002, as amended.
8. Part 2 of the IET Wiring Regulations contains the Definitions.
9. Manual Handling Operation Regulations.
10. PUWER 1998 Provision and Use of Work Equipment Regulations.

Section 2 Health and Safety

1. Any of the tools listed in Section 2, such as a hacksaw. A risk associated with using a hacksaw would include cuts and lacerations. Another example is a hammer: this could cause injury to hands and fingers if used carelessly and could also damage equipment if not used for its intended purpose, for example, using it to straighten a switch that is not quite level.

2 The insulation on tools designed for electrical work is to protect the user from electric shock.
3 For the purposes of your examinations, the key should be kept in your pocket. If you don't have a pocket you should find a way to keep the key with you at all times.
4 False: a risk assessment should be periodically reviewed throughout the entire duration of the work being carried out.
5 Wire strippers. You may encounter experienced electricians using various other tools, such as side cutters or pliers. However, these are more likely to damage the conductor and cause it to break when terminating.

Section 3 Generation and Transmission

1 Coal, gas, oil, nuclear, petrol/diesel and anything else that will be depleted if we continue using it as a source of energy.
2 Electrical Safety Quality and Continuity Regulations (ESQCR) 2002, as amended.
3 Electricity (alternating current) can be transmitted much more efficiently when the voltage is increased.
4 Distribution transformers are generally found close to areas where the electricity is being used.
5 Electricity in the UK should be at 50 Hz.
6 Electricity is usually charged per kilowatt hour (kWh).
7 There are many aspects of the National Grid but one of the greatest, mentioned in Section 3, is that the entire UK's power supply is connected together so that if, for some reason, one power station fails, the towns and cities in that area will still have power.
8 Step up transformers increase voltage.
9 Step down transformers reduce voltage.
10 Most domestic homes in the UK will have a 230 V supply although larger homes may have a larger supply.

Section 4 Supply

1. Star connection and delta connection.
2. Protective Earthed Neutral (a combined protective and neutral conductor in the cable supplying an installation).
3. Information must include:
    (a) nominal voltage;
    (b) nature of current and frequency;
    (c) prospective short-circuit current;
    (d) external earth fault loop impedance;
    (e) suitability for maximum demand; and
    (f) type and rating of overcurrent device at the origin.
4. Single-core, insulated and sheathed cable.
5. Regulation 421.1.201.
6. Any suitable example given in Regulation 314.1.
7. Eight.
8. Regulation 536.4.5.
9. 32 A.

Section 5 Protection and Isolation

1. An overcurrent occurring in a circuit that is electrically sound.
2. Asbestos.
3. The serviceable short-circuit current rating.
4. The approximate disconnection times are:
    (a) 1000 seconds or 17 minutes.
    (b) 500 seconds or 8.5 minutes.
    (c) 100 seconds or 1.5 minutes.
    (d) 250 seconds or 4 minutes.
5. 30 mA.
6. Chapter 53.
7. Regulation 537.2.5: Provision shall be made for securing off-load isolating devices against unwanted or unintentional opening. This may be achieved, for example, by locating the device in a lockable space or lockable enclosure or by padlocking. Alternatively, the off-load device may be interlocked with a load-breaking one.

8   Examples may include:
    (a) light switches;
    (b) time clocks;
    (c) PIR sensors; and
    (d) contactors.
9   The machine could suddenly re-start should the power come back on following a power cut.
10 A contactor or direct-on-line starter unit having start and stop buttons (see 5.11 for details).

Section 6   Earthing and Bonding

1   No, the earthing conductor is not a live conductor.
2   Terre Terre (TT) systems are usually the first choice for providing a means of earthing if the electricity supplier cannot provide a means of earthing.
3   In order for a protective device, such as a circuit-breaker, to operate, the fault current must have a good path to earth returning to the neutral conductor at the distribution transformer so that it completes the earth fault path.
4   Earth electrode.
5   As below:
    (a) Terre Neutral-Separated: the earth is a separate conductor connected to the neutral terminal at the distribution transformer.
    (b) Terre Neutral-Connected-Separated: the earth and neutral conductors come into the property as one conductor and are separated at the consumer's installation.
    (c) Terre Terre: this is where the general mass of the Earth is used to provide a path for the fault current to travel back to the neutral at the distribution transformer.
6   Exposed-conductive-part: this is part of a piece of electrical equipment or installation.
7   Extraneous-conductive-part: this is not part of the electrical installation but it is in contact with the general mass of Earth and may provide a path for fault current to travel back to the neutral at the distribution transformer.

8 During a fault to earth.

9 Close to zero.

10 It is required so that if one of the connections to an extraneous-conductive-part becomes damaged the conductor will still provide a path for current to flow to the neutral at the distribution transformer.

Section 7 Cable calculation

Task one

$I_b = P/V$, so $I_b = 9000/230$

$I_b = 39.13$ A

Select I_n:

$I_n \geq 39.13$ (I_n = BS 60898 40 A (the next available size up from 39.13))

So $I_b \leq I_n$

Rating factors

$C_a = 0.94$

$C_i = 1$ (using Table F6 of the *On-Site Guide* and Table 4D5 of the IET Wiring Regulations) can be used

$C_f = 1$ (as the device being used is not a semi-enclosed fuse)

C_c = not applicable

C_d = not applicable

C_g = not applicable

C_s = not applicable

Select the right size cable (using Table 4D5 of the IET Wiring Regulations)

$I_t \geq \dfrac{I_n}{C_a \times C_i \times C_f}$

$I_t \geq \dfrac{40}{0.94 \times 1 \times 1}$

$I_t = 42.55$ A

According to Table 4D5 of the IET Wiring Regulations, a 10 mm² cable will be adequate to carry the design current when installed using method 100.

Check that the Voltage drop is within the permitted values given in Table 4Ab of the IET Wiring Regulations.

$$Voltage\ drop = \frac{(mV/A/m \times I_b \times L)}{1000}$$

$$Voltage\ drop = \frac{(4.4 \times 39.13 \times 6.3)}{1000}$$

Volt drop = 1.08 V

As this is within the permissible value for volt drop (5 % for other uses) this suggests that the cable is suitable for the job.

Task two

$I_b = P/V$, so $I_b = 500/230$

$I_b = 2.17$ A

Select I_n

$I_n \geq I_b$

I_n = BS EN 60898 6 A (the next available size up from 2.17)

$I_n \geq 39.13$ A

Rating factors

$C_a = 1$

C_i = not applicable

$C_f = 1$ (as the device being used is not a semi enclosed fuse)

C_c = not applicable

$C_d = 1.03$

C_g = not applicable

$C_s = 1$

Select the right size cable (Table 4D5 of the IET Wiring Regulations)

$$I_t \geq \frac{I_n}{C_a \times C_f \times C_d \times C_s}$$

$$I_t \geq \frac{6}{1 \times 1 \times 1.03 \times 1}$$

$I_t = 5.82$ A

According to Table 4D4A of the IET Wiring Regulations, a 1.5 mm² cable will be adequate to carry the design current when installed using reference method D.

Check that the Voltage drop is within the permitted values given in Table 4D5 of the IET Wiring Regulations.

$$Voltage\ drop = \frac{(mV/A/m \times I_b \times L)}{1000}$$

$$Voltage\ drop = \frac{(29 \times 2.17 \times 20)}{1000}$$

Volt drop = 1.26 V

As this is within the permissible value for volt drop (5 % for other uses) this suggests that the cable is suitable for the job.

Section 8   Final Circuits

1. Fused and unfused.
2. 30 A or 32 A.
3. A fused connection unit
4. 559.5.1.201 states: 'A ceiling rose or lampholder shall not be installed in any circuit operating at a voltage normally exceeding 250 volts'.
5. Regulation 559.9
6. As below:
   (a) Off.
   (b) On.
   (c) Off.
   (d) On.

7 Extra low voltage to reduce the severity of electric shock and separation of the electrical supply from earth.
8 PELV does not separate the circuit from earth.
9 Part 7 section 701.
10 Control systems.

Section 9 Inspection and Testing

1 An insulation resistance test will allow the tester to identify whether or not the insulation between the conductors is sufficient.
2 Highest.
3 The line and earth (cpc) must be connected together (linked) at the consumer unit.
4 Maximum Z_s values can be found in Appendix B.
5 The switches may need to be operated in order to obtain a reading.
6 The recommended length is 2 mm and they should be no more than 4 mm.

Section 10 Fault Finding

1 If a fault cannot be rectified the circuit must be left in a safe condition until the fault can be rectified. For example, if additional material is required from the wholesaler the electrician should make the installation safe before leaving.
2 A low resistance would indicate a satisfactory earth fault path.
3 The first logical step would be to split the circuit into two and retest each part of the circuit so that the part with the fault can be found (don't get confused with identifying and locating the fault).
4 Once a fault has been rectified the sequence of tests must be carried out as they would on a new installation or any part of the installation that has been affected by the fault finding process.

Section 11 Common Calculations

1 As below:
   (a) 12
   (b) 9
   (c) 3
2 Watts
3 $I \times R = V$
4 2 A
5 As below:
   (a) 9.5 Ω
   (b) 51 Ω
   (c) 15 Ω
6 (a) 2.1 Ω
   (b) 0.62 Ω

Section 12 Diversity

1 The protective device should be rated at 15.5 A or the next available higher rating, for example, a 16 A circuit-breaker to BS EN 60898.
2 As below:
   (a) The installer may over-engineer the installation at additional cost.
   (b) The customer would have an expensive installation that will never be used to its full capacity.
3 100 % full load (f.l.) of the largest appliance + 80 % f.l. of the second largest appliance + 60 % f.l. of remaining appliances.
4 Any electrical appliance that is designed to run continuously at full load, such as a shower.
5 Guidance provided in the IET's *On-Site Guide* asks installers to assume that there will be a demand of 100 W per lampholder on a lighting circuit.
6 The working should look something like this:

$I_n \geq I_b$

$I_b = \dfrac{P}{V}$

The total P is broken into the individual parts of the cooking appliance:

P = (2 X 2,000) + (2 X 750) + 3,000

$$I_b = \frac{8,500}{230}$$

I_b = 37 A

As stated in the question, this cooking appliance has a socket-outlet included, so an additional 5 A is added to I_b.

37 + 5 = 43

I_b = 43 A

Now that the total I_b has been worked out, we can apply diversity.

10 A is taken from I_b

43 − 10 = 33

I_b = 33 A

Only 30 % of the remaining 33 A is taken.

33 × 30 % = 9.9 A

I_b = 9.9 A

At this stage you can add the 5 A for the socket-outlet and the 10 A that was taken at the beginning:

I_b = 9.9 + 5 + 10

I_b = 24.9 A

I_n ≥ 24.9 A

The circuit-breaker to be installed for this circuit should have a rating of 30 A or 32 A.

Index

A

abbreviations	Table 1.3; Table 1.4
acceptable risk	2.2.3; Table 2.2
additions and alterations to an installation	8.2.1
ambient temperature	7.2.2; 7.3.2
appliance demands	12.3
awarding bodies	1.5.2

B

bonding	Section 6
breaking capacity	5.1
British Standards	Table 1.1
buried cables	7.2.2

C

cable ladders	Table 1.5
cables	
calculations	Section 7
current-carrying capacity	1.7.2
grouping	7.2.2
installing	8.2.3
lighting circuits	8.3.2
rating	1.7.3; 7.2; 7.3
routes	8.2.4
terminology	Table 1.5
types	1.5.1; Table 1.2
voltage drop	7.4
cable trays	Table 1.5
cartridge fuses	5.1.2
ceiling roses	8.3.2
CENELEC (European Committee for Electrotechnical Standardization)	Table 1.1
circuit-breakers	5.1.5
standards	Table 5.1
circuit design: *see final circuits; see lighting circuits*	
circuit protective conductors (cpc)	6.1.1
continuity testing	9.4.2; 9.4.3
ring final circuits	8.5.1; 8.5.2
circuit protective devices: *see protective devices*	
competent person scheme	Table 1.3
conduits	Table 1.5
consumer's tails	4.6
consumer units	4.7
heights	8.2.3
installing	8.2.1; 8.2.2

Index

ways and modules	4.7.1
continuity testing	9.4.1; 9.4.2
cpc: *see circuit protective conductors (cpc)*	
cross polarity	10.1.3; 10.2.5
cross sectional areas (CSAs)	8.5.2; 8.5.4
current-carrying capacity	1.7.2
current causing effective disconnection	5.1
current rating	
of cables and equipment	1.7.2; 7.2; 7.3
of supply	4.3

D

dead tests	9.4
design current	7.2.1; 7.3.1; 7.3.2
disconnection current	5.1
discrimination: *see selectivity*	
distribution board (DB)	4.7
distribution circuits	4.7
distribution network operators (DNOs)	4.3.1; 6.1
distribution networks	3.3.3
distribution transformers	3.3.3
diversity	Section 12
DNOs (distribution network operators)	4.3.1; 6.1

E

earth electrodes	4.4.1; 4.4.3; 6.1.4
earth fault loop	6.2.4
earth fault loop impedance (EFLI)	1.7.4; 8.2
testing	9.5.2; 9.5.3
earth faults	5.1; 6.1.2; 6.2.3; 10.2.2
earthing	Section 6
arrangements	4.4; 6.1.4
protective equipotential bonding	6.2
responsibility for	6.1
supplementary equipotential bonding	6.3
EFLI: *see earth fault loop impedance (EFLI)*	
electrical isolation: *see isolation*	
Electricity at Work Regulations 1989 (EWR)	9.1.1
electricity generation	Section 3
electricity meters	3.4
Electricity Safety, Quality and Continuity Regulations (ESQCR)	1.5.3
electricity transmission	3.3
electric shock	6.2.3; 6.2.4
emergency switching	5.9
environmental conditions	8.3.3
equipotential bonding: *see protective equipotential bonding conductors*	
ESQCR (Electricity Safety, Quality and Continuity Regulations)	1.5.3
European Committee for Electrotechnical Standardization (CENELEC)	Table 1.1

Index

European Standards	1.2; Table 1.1	functional switching	5.10
exposed-conductive-part	Table 1.2	fuse board	4.7
extra low voltage	1.5.1; 8.8; 8.9; 8.10; Table 1.2	fuse carrier	4.3.1
extraneous-conductive-part	Table 1.2	fuses	
		and cable rating	7.2.2
		cartridge	5.1.2

F

fault finding	Section 10	high rupturing capacity (HRC)	5.1.1
FELV (Functional Extra-Low Voltage systems)	8.10	plug-top	5.1.3
		semi-enclosed	5.1.4
final circuits: *(see also radial final circuits); (see also ring final circuits)*	4.7; Section 8	service	4.3.1
		standards	Table 5.1

G

additions and alterations to an installation	8.2.1	generators	Section 3
cable routes	8.2.4	Guidance Note GS38	2.3.1

H

consumer unit arrangements	8.2.2	hand tools	2.4; Table 2.3
designing and installing new circuits	8.2	health and safety: *(see also electric shock)*	Section 2
Functional Extra-Low Voltage systems (FELV)	8.10	Health and Safety at Work Act 1974 (HSWA)	2.1
general installation practices	8.1	Health and Safety Executive	1.5.3; Table 1.4
heights of switches, socket-outlets and consumer units	8.2.3	heights of fittings	8.2.3
		high resistance	10.1.2; 10.1.5; 10.2.4
luminaires and lighting installations	8.3	high rupturing capacity (HRC) fuses	5.1.1
Protective Extra-Low Voltage systems (PELV)	8.9	hydro generation	3.1.2
Separated Extra-Low Voltage systems (SELV)	8.8	**I**	
shower circuits	8.7	I_{CN} rating	5.1.5
spurs	8.6	I_{CS} rating	5.1.5
Functional Extra-Low Voltage systems (FELV)	8.10		

Index

IEC (International Electrotechnical Commission)	Table 1.1
IET Wiring Regulations	1.1
initial inspection	9.2; Table 9.1
inspection: *(see also testing)*	Section 9
initial inspection	9.1.3; 9.2
legal aspects	9.1.1
responsibilities	9.1.2
installation	1.7.1; 8.1; 8.2
location symbols	Table B.10
institutions	1.5.2; Table 1.3
insulation resistance	
minimum values	9.4.4; Table 9.2
testing	9.4.4
integration of devices and components	5.4
interconnection of conductors	10.1.4; 10.2.6
intermediate light switches	8.3.2
International Electrotechnical Commission (IEC)	Table 1.1
isolating transformers	8.8.1; 8.9.1
isolation: *(see also switching)*	5.5
for mechanical maintenance	5.8
for particular installation types	5.7; Table 5.3
symbols	Table B.7; Table B.8

isolators	
compared with switches	5.6
isolation switch (mains)	4.5; 4.7

J

JPEL/64 Committee	1.3

K

kitchen appliance demands	12.3

L

legal aspects	1.6; 9.1.1
lighting circuits	8.3.2; 9.4.2
lighting installations	8.3
environmental conditions	8.3.3
regulations and guidance	8.3.1
selection of equipment	8.3.2; 8.3.3
light switches	8.3.2
link method	9.4.2
live conductor arrangements	4.2; Table 4.1
live tests	9.5
long lead method	9.4.2
luminaires	8.3; 8.3.2

M

main switch	4.7
maintenance, switching off for	5.8
materials	Table 1.5
maximum demand: *(see also diversity)*	4.3
mechanical maintenance	5.8

Index

meters	3.4; Table B.9	policies	1.6
meter tails	4.6	potential difference	6.2.2
method statements	2.2.2	power distribution	3.3.3
miniature circuit-breakers (mcb)	5.1.5	power generation	Section 3
motor circuits	5.11	power law	11.2
moulded case circuit-breakers (mccb)	5.1.5	power sources	3.1
multiples and sub-multiples of quantities	Table B.3	power stations	3.1.1
		power supply characteristics	3.2
		power tools	2.4; Table 2.4

N

National Grid	3.3	power transmission	3.3
neutral to earth faults	5.1; 8.2.1	pre-determined values	1.7
nominal rating	5.1	prospective fault currents (PFC)	5.1; 5.1.5; 5.3
		testing	9.5.4

O

Ohm's law	11.2	protective conductor current	5.1
open circuits	10.2.3	protective conductors: *see circuit protective conductors (cpc)*	
organisations	1.5.2; Table 1.3		
overcurrent	5.1	protective devices: *(see also circuit-breakers); (see also fuses)*	4.7; 5.1
overload current	5.1; 10.1.6	current rating	5.1; 7.2; 7.3
overload protection	7.2.2	selecting	5.2

P

		standards	Table 5.1; Table 5.2
parallel circuits	11.3.2		
PELV (Protective Extra-Low Voltage systems)	8.9	protective earthed neutral (PEN) conductor	4.4.1; 6.1.4
PEN (protective earthed neutral) conductor	4.4.1; 6.1.4	protective equipotential bonding conductors	6.2
		continuity testing	9.4.1
phase sequence testing	9.5.5	Protective Extra-Low Voltage systems (PELV)	8.9
plug-top fuses	5.1.3		
PME (protective multiple earth)	4.4.1; 6.1.4	protective multiple earth (PME)	4.4.1; 6.1.4
polarity testing	9.4.5; 9.5.1		

Index

Q
qualifications	1.3.2

R
radial final circuits	8.4
continuity testing	9.4.2
maximum area served	8.4.5
maximum number of socket-outlets	8.4.3
spurs	8.6
rated short-circuit capacity	5.1
rating factors	1.7.3; 7.2.2
residual current breaker with overload (RCBO)	5.1.7
residual current devices (RCDs)	5.1.6; 8.2.2
testing	9.5.6
resistors	11.3
rewirable fuses: *see semi-enclosed fuses*	
ring final circuits	8.5
circuit protective conductors (cpc)	8.5.1
continuity testing	9.4.3
diversity	12.3
fault finding	10.2.7
maximum area served	8.5.4
maximum number of socket-outlets	8.5.3
spurs	8.6
risk assessments	2.2; 9.1.2

S
safe isolation procedure	2.3
selectivity	5.3
semi-enclosed fuses	5.1.4; 7.2.2
Separated Extra-Low Voltage systems (SELV)	8.8
series circuits	11.3.1
service fuses	4.3.1
short circuits	5.1; 10.1.1; 10.2.2
shower circuits	8.7
single-phase supply	4.2
SI units	1.4; Table B.2; Table B.4
socket-outlets: *(see also radial final circuits); (see also ring final circuits)*	
heights	8.2.3; 8.4.4
maximum number	8.4.3; 8.5.3
spurs	8.6
soil resistivity	7.2.2
solar power	3.1.3
special locations	Annex A
split-way boards	4.7.1
spurs	8.6
standards	Table 1.1
power supply characteristics	3.2
protective devices	Table 5.1; Table 5.2
statutory requirements	1.6
substations	3.3.2
supplementary equipotential bonding	6.3
supply intake	4.1; 4.5

Index

current rating	4.3
earthing arrangements	4.4
live conductor arrangements	4.2; Table 4.1
responsibility	4.5
tails	4.6
supply rating	4.3
switches	
compared with isolators	5.6
heights	8.2.3; 8.4.4
lighting circuits	8.3.2
symbols	Table B.6 Table B.8
switching: *(see also isolation)*	5.5
emergency	5.9
functional	5.10
undervoltage protection	5.11
symbols	1.4; Annex B

T

tails	4.6
terminology	1.5.1; Table 1.5; Table 1.6
testing	Section 9
dead tests	9.4
initial inspection	9.1.3; 9.2
legal aspects	9.1.1
live tests	9.5
responsibilities	9.1.2
thermal insulation	7.2.2
three-phase supply	4.2
three-plate method	8.3.2
TN-C-S systems	4.4.1; 6.1; 6.1.4
TN-S systems	4.4.2; 6.1.4
tool safety	2.4
touch voltage	6.2.3; 6.2.4
transmission networks	3.3.2
transposition	11.1
triangle method	11.2
trunking	Table 1.5
TT system	4.4.3; 4.7
TT systems	6.1.1; 6.1.4
two-plate method	8.3.2
two-way light switches	8.3.2

U

undervoltage protection	5.11

V

verification of installation	9.1.3
visual inspection	10.2.1
voltage drop	7.4

W

wandering lead method	9.4.2
wind generation	3.1.2
wiring diagrams	Table B.5

ENDORSED BY

Online training for BS 7671:2018

With the IET's interactive online training from the IET Academy, there's no need to take time away from the job to train for your C&G 2382:18 (or equivalent) qualification.

Book a full or Update course and get:

- Courses that have been endorsed by C&G, ensuring the best quality content
- Learning at your own pace with 12 months access to all content
- Online/offline access to suit you, through our app
- Three full practice exams to fully prepare you for the real thing
- A certificate of completion to present to your local exam centre to sit your C&G 2382:18 exam

Find out more and book your course now at:
www.theiet.org/academy-regsa